THE CONCISE LEXICON OF ENVIRONMENTAL TERMS

WITH A LIST OF SCIENTIFIC ABBREVIATIONS
AND GLOSSARY OF ACRONYMS

EDITORS

Professor Malcolm Grant
Department of Land Economy, Cambridge University

Richard Hawkins
MA Barrister FRSA FRGS F INST WM

JOHN WILEY & SONS
Chichester • New York • Brisbane • Toronto • Singapore

First published in Great Britain in 1995 by United Kingdom Environmental Law Association in association with McKenna & Co

This edition published by John Wiley & Sons Ltd
Baffins Lane, Chichester
West Sussex, PO19 1UD England

National 01243 779777
International +44 1243 779777

Published in North America by John Wiley & Sons, Inc
7222 Commerce Center Drive, Colorado Springs CO 80919 USA

Other Wiley Editorial Offices

John Wiley & Sons, Inc., 605 Third Avenue,
New York, NY 10158-0012, USA

Jacaranda Wiley Ltd, 33 Park Road, Milton,
Queensland 4064, Australia

John Wiley & Sons (Canada) Ltd, 22 Worcester Road,
Rexdale, Ontario M9W ILI, Canada

John Wiley & Sons (SEA) Plc Ltd, 37 Jalan Pemimpin #05-04,
Block B, Union Industrial Building, Singapore 2057

British Library Cataloguing in Publication Data:

A CIP catalogue record for this book is available from the British Library

ISBN 0-471-96357-7

Typeset in Stone Serif by Tony & Penny Mills, Beccles, Suffolk

Printed and bound in Great Britain by Butler & Tanner Ltd, Frome and London

CONTENTS

ACKNOWLEDGEMENTS

For their help in preparing this Lexicon, the editors are hugely indebted to many contributors, although any errors and omissions are solely the responsibility of the joint editors. The many contributors include:

Ross Fairley and Owen Lomas (of Allen & Overy); Trevor Adams (of Ashurst Morris Crisp); Patricia Henton, Stephen Last, Stephen Rintoul, Howard Robinson and Peter Young (of Aspinwall & Co); Richard Burnett–Hall (of Bristows Cooke & Carpmael); Steve Yerby (of C E Heath Insurance Brokers); Trevor Winterbottom (of Cambridge City Council); Tricia McLoughlin and Sam Johnston (of Cambridge University, Department of Land Economy); Sue Cooper and Clare Lodge (of Cornard Tye House); Malcolm Forster and Peter Kavanagh (of Freshfields); Ruth Grinbergs (of Howard Kennedy); Jane Bickerstaffe (of INCPEN); Julie Hill, Robert Huxford, Ken Marchant, Diana Maxwell, Michael Philpott, Chris Tunaley, Philip Watts (of The Institution of Civil Engineers Waste Management Forum); Keith Knox (of Knox Associates); Daniel Lawrence (of Lovell White Durrant); Adrian Blomfield, Pamela Castle, Helen Harrison, Ann Peirson–Hills and Paul Sheridan (of McKenna & Co); Kate Alexander, Denis Costello, Christine Cowley, Fiona McMurray, Rebecca Rusling (of Mills & Reeve); Donald Reid (of Morton Frazer Milligan WS); Pandora Rene (of North Wales Waste Management Group); Nicholas Jones (of Overbury, Steward and Eaton); Andrew Wiseman (of Shindler & Co); Robert Evans, James Fitzgerald, Kathy Mylrea, William Rawley, Stephen Tromans and Carlos Vilhena (of Simmons & Simmons); David Cuckson (of Stephenson & Harwood); Penny Jewkes (of University of East Anglia); Ben Pontin (of University of Southampton, Faculty of Law); Rob Stevens (of University of Southampton, Urban Research Policy Unit); Denise Artis (of University of Central Lancashire, Department of Law); Robert McCracken (of 2 Harcourt Buildings); Dr Christina Hill; Dr Wendy Le–Las.

PREFACE

The principal objective for the research, editing and collation of this Lexicon has been to provide for not only environmental lawyers but also for all professionals, bankers, planners, chartered surveyors, environmental health officers, architects, writers and journalists, campaigners, community groups, local authority councillors and officers and, above all, for the generally well-informed layman and woman a readily available quick guide to certain environmental terms which are in common use. The editors have also tried to provide useful indications for those readers who wish to obtain more detailed source material.

The joint editors are particularly grateful for the work of the collating editors, Pamela Castle, Malcolm Forster, Nicholas Jones and particularly Stephen Tromans. The work of the individual researchers is also acknowledged on the preceding page. We would also like to thank Penny Mills for the book and page design, achieved often under considerable pressure.

The editors would also like to single out the work of Ann Peirson-Hills and the McKenna & Co database team who were the real engine room of this publication.

UKELA is again grateful to McKenna & Co and also to Aspinwall & Company for financially underwriting this project and for the generous support in management time that all at Mitre House and Walford Manor have provided.

No work can be complete, particularly as wide a selection as contained in the following pages and which has been unavoidably a matter of considered subjective and selective judgement. We can only quote, and not too lamely, Einstein whose remark prefaces Chapter 1 of the Royal Commission on Environmental Pollution's 1984 10th Report, namely 'The environment is everything that isn't me.'

Accordingly, the editors would be appreciative of any suggestions for additions or deletions and corrigenda for a further edition.

MALCOLM GRANT

Department of Land Economy,
University of Cambridge, 19 Silver Street,
Cambridge CB3 9EP
(F) 01223 337132

RICHARD HAWKINS

Cornard Tye House, Sudbury,
Suffolk CO10 OQA
(F) 01787 377705

May 1995

ABOUT UKELA

The United Kingdom Environmental Law Association is a company limited by guarantee and registered as a charity. It was formed in 1987 at a meeting at the University of Southampton for all those interested, not necessarily lawyers, in the development of environmental law and practice in the EU and UK. Over the last 8 years it has grown from initial membership of 7 to over 1,400 members who pay £15 per year, £40 per year corporate membership, £5 student membership. All enquiries, please, to the Membership Secretary, Professor Malcolm Forster, Freshfields, 65 Fleet Street, London, EC4Y 1HS (T) 0171 936 4000 (F) 0171 832 7001, or the Secretary, Dr Christina Hill, Honeycroft House, Pangbourne Road, Upper Basildon, Berkshire, RG8 8LP (T and F) 01491 671631.

UKELA's publication is *Environmental Law*. This publication is distributed quarterly free to all members. All enquiries with regard to the journal (which is also available on subscription to non-members) should be addressed to the Editor: Owen Lomas, Allen & Overy, One New Change, London, EC4M 9QQ (T) 0171 330 3000 (F) 0171 330 9999.

UKELA holds regular meetings at which UK and international contemporary environmental practical legal issues are discussed, often with distinguished speakers. There is an annual prize moot and also a weekend conference which is hosted at a different regional centre each year.

SCIENTIFIC ABBREVIATIONS

SI Units

Quantity	**Unit**	**Symbol**
Length	Metre	m
Mass	Kilogram	kg
Time	Second	s
Area	Metre squared	m^2
Volume	Metre cubed	m^3
Volume (fluids)	Litre	l
Velocity	Metres per second	ms^{-1}
Force	Newton	N
Energy	Joules (mN)	J
Power	Watt (J/S)	W
Density	Kilogram per cubic metre	kgm^{-3}
Pressure	Pascal (N/m^2)	Pa
Pressure (for meteorology)	Millibar	mb
Electric current	Ampere	A
Potential difference	Volts	V
Electrical resistance	Ohm	Ω
Electrical conductivity	Siemens (Ω^{-1})	S
Frequency	Hertz	Hz
Temperature	Kelvin	K
Noise	Decibel	dB
Radioactive activity	Becquerel	Bq
Radiation dose	Gray (J/kg)	Gy
Acidity	pH	–

Multiples and Sub-multiples

Name	**Value**	**Prefix**
Pico	10^{-12}	p
Nano	10^{-9}	n
Micro	10^{-6}	μ
Milli	10^{-3}	m
Kilo	10^{3}	k
Mega	10^{6}	M
Giga	10^{9}	G
Tera	10^{12}	T

Multiples and Sub-multiples

one billion	10^9	'giga'	G	eg	gigahertz	(GH_z)
one million	10^6	'mega'	M	eg	megawatt	(MW)
one thousand	10^3	'kilo'	K	eg	kilojoule	(kJ)
one hundred	10^2	'hecto'	H	eg	hectowatt	(hW)
ten	10^1	'deca'	D	eg	dekametre	(dam)
unity						
one tenth	10^{-1}	'deci'	d	eg	decilitre	(d)
one hundredth	10^{-2}	'centi'	c	eg	centimetre	(cm)
one thousandth	10^{-3}	'milli'	m	eg	millivolt	(mV)
one millionth	10^{-6}	'micro'	µ	eg	microgram	(µg)
one billionth	10^{-9}	'nano	n	eg	nanosecond	(nsec)
one trillionth	10^{-12}	'pico'	pp	eg	picogram	(pg)

Other Commonly Used Units

Quantity	Unit	Symbol	Equivalent
Area	Hectare	Ha	10,000 m^2
Temperature	Degrees Celsius	°C	K+273
	Degrees Centigrade	°C	As Celsius
	Degrees Fahrenheit	°F	(°C x 1·8) + 32
Mass	Tonnes	t	1,000 kg

Abbreviation	Term
AOD	Above Ordnance Datum
BH	Borehole
BOD	Biochemical Oxygen Demand
BOD_5	5 Day BOD
Btu	British thermal unit
CFC	Chlorofluorocarbon
CH_4	Methane
CO_2	Carbon Dioxide
COD	Chemical Oxygen Demand
CQA	Construction Quality Assurance
dB(A)	'A' weighted noise measurement
FFT	Fast Fourier Transformation
GAC	Granular Activated Carbon
GC	Gas Chromatography
HDPE	High Density Polyethylene
H_2S	Hydrogen Sulphide

IRS	Infra–red Spectroscopy
L_{AEQT}	'A' weighted equivalent noise level for time 'T'
LC_{50}	Lethal concentration of a substance that will kill 50% of all organisms exposed to that concentration
LD_{50}	Lethal dose of a substance required to kill 50% of all organisms exposed to that dose
LFG	Landfill Gas
L_n	Level that noise exceeds for n% of time
MC	Moisture Content
MEL	Maximum Exposure Limit
MS	Mass Spectrometry
NNI	Noise & Number Index
NO_x	Nitrogen oxides
O_2	Oxygen
PAH	Polyaromatic Hydrocarbons
PCB	Polychlorinated Biphenyl
PNAH	Polynuclear Aromatic Hydrocarbons
PNdB	Perceived noise level
ppa	Peak Particle Acceleration
ppv	Peak Particle Velocity
SPL	Sound Pressure Level
SPT	Standard Penetration Test
SS	Suspended Solids
SWL	Sound Power Level
T	Radioactive half life
TOC	Total Organic Carbon
TP	Trial Pit
TPH	Total Petroleum Hydrocarbons
VOC	Volatile Organic Carbon

What Scientific Abbreviations Can Really Mean

Municipal Waste to Energy Plant (MWTE)

Dioxin standards should be at a concentration of about 0·1 ng/m^3; this of course means nanograms per cubic metre of gas. What this means in practice, is that 1 gram of dioxin is contained in 10,000,000,000 cubic metres of cooled stack gas. Let us place this in context.

The volume of Britain's largest lake, Loch Ness, is approximately 7,000,000,000m^3 so three-quarters of a 3 gram sugar cube dissolved in the whole of Loch Ness equates to the concentration of dioxins in the stack gases leaving the top of the chimney. Thus other uncontrolled combustion sources of dioxins such as treated wood waste, straw, tyres, coal, 5th November bonfires and fireworks and landfill gas may now require equally serious attention.

SOURCE: Professor Andrew Porteous,
The Open University,
at IWM 1994 Conference Workshop

ENVIRONMENTAL ACRONYMS IN GENERAL USE

Acronyms

AAI	Area of Archaeological Importance
ACBE	Advisory Committee on Business and the Environment
ACHS	Advisory Committee on Hazardous Substances
ACOP	Approved Code of Practice
ACRE	Advisory Committee on Releases to the Environment
ADAS	Agricultural Development and Advisory Service
ADI	Acceptable Daily Intake
ADR	1 Accord Européen Relatif au Transport International des Marchandises Dangereuses par Route (European Agreement on the International Carriage of Dangerous Goods by Road) or 2 Alternative Dispute Resolution
AEC	Association of Environmental Consultancies
AGEMA	Advisory Group on Eco-management and Audit
AONB	Area of Outstanding Natural Beauty
BAT	Best Available Technology
BATNEEC	Best Available Techniques/Technology Not Entailing Excessive Cost
BEO	Best Environmental Option
BOD	Biological/Biochemical Oxygen Demand
BPEO	Best Practicable Environmental Option
BPM	Best Practicable Means
BREEAM	Building Research Establishment Environmental Assessment Method
BS	British Standard
BS 5750	British Standard 5750 on Quality Systems
BS 7750	British Standard 7750 on a Specification for Environmental Management Systems
BTKNEEC	Best Technical Knowledge Not Entailing Excessive Costs
CAP	Common Agricultural Policy
CBA	Cost-Benefit Analysis
CC	Countryside Commission

CCA	Compliance Cost Assessment
CCGT	Combined Cycle Gas Turbine
CCW	Countryside Council for Wales
CERCLA (US)	Comprehensive Environmental Response, Compensation and Liability Act
CERES	Coalition for Environmentally Responsible Economies
CFCs	Chlorofluorocarbons
CGL (US)	Comprehensive General Liability
CHIP	Chemicals (Hazard Information and Packaging) Regulations 1994
CHP	Combined Heat and Power
CIA	Chemical Industries Association
CIGN	Chief Inspector's Guidance Notes
CIMAH	Control of Industrial Major Accident Hazard Regulations 1984
CITES	Convention on International Trade in Endangered Species
CIWEM	Chartered Institute of Water and Environmental Management
CLC	International Convention on Civil Liability for Oil Pollution Damage 1969
CLEUD	Certificate of Existing Lawful Use or Development
CLOPUD	Certificate of Lawful Proposed Use or Development
COD	Chemical Oxygen Demand
COPA	Control of Pollution Act 1974
COSHH	Control of Substances Hazardous to Health Regulations 1994
COTC	Certificate of Technical Competence
CPL	Classification, Packaging and Labelling of Dangerous Substances Regulations 1984
CPRE	Council for the Protection of Rural England
CRI	Chemical Release Inventory
CRISTAL	Contract Regarding a Supplement to Tanker Liability for Oil Pollution
DDT	Dichloro Diphenyl Trichloroethane
DLG	Derelict Land Grant
DNA	Deoxyribonucleic Acid
DOE	Department of the Environment
DSD	Duales System Deutschland
DTI	Department of Trade and Industry

DTp	Department of Transport
DWF	Dry Weather Flow
EAP	Environmental Action Programme
EARA	Environmental Auditors Registration Association
EC	European Communities
ECJ	European Court of Justice
ECOSOC	Economic and Social Committee
EEA	European Environment Agency
EEB	European Environmental Bureau
EEC	European Economic Community
EH	English Heritage
EHO	Environmental Health Officer
EIA	Environmental Impact Assessment
EIL	Environmental Impairment Liability
EINECS	European Inventory of Existing Commercial Chemical Substances
EIONET	Environmental Information and Observation Network
ELINCS	European List of Notified Chemical Substances
EMAS	Eco-management and Auditing Scheme
EMF	Electromagnetic Fields
EMR	Electromagnetic Radiation
ENERO	European Network of Environmental Research Organisation
EPA 1990	Environmental Protection Act 1990
EPA (US)	Environmental Protection Agency
EQO	Environmental Quality Objectives
EQS	Environmental Quality Standards
ERRA	European Recovery and Recycling Association
ESA	Environmentally Sensitive Area
EST	Energy Savings Trust
ETBPP	Environmental Technology Best Practice Programme
ETIS	Environment Technology Innovation Scheme
ETSU	Energy Technology Support Unit
EU	European Union
EWC	European Waste Catalogue
FEPA	Food and Environment Protection Act 1985
FID	Flame Ionisation Detector
FoE	Friends of the Earth

GATT	General Agreement on Tariffs and Trade
GC/MS	Gas Chromatography/Mass Spectrometry
GDO	Town and Country Planning General Development Order 1988
GEF	Global Environment Facility
GMO	Genetically Modified Organism
GPR	Ground Penetrating Radar
HBFCs	Hydrobromofluorocarbons
HCFCs	Hydrochlorofluorocarbons
HMIP	Her Majesty's Inspectorate of Pollution
HMIPI (Scotland)	Her Majesty's Industrial Pollution Inspectorate
HSC	Health and Safety Commission
HSE	Health and Safety Executive
ICRCL	Interdepartmental Committee on the Redevelopment of Contaminated Land
ICRP	International Commission on Radiological Protection
IEA	Institute of Environmental Assessment
IEHO	Institute of Environmental Health Officers
IEM	Institute of Environmental Management
IMO	International Maritime Organisation
INCPEN	Industry Council for Packaging and Environment
IPC	Integrated Pollution Control
IPPC	Integrated Pollution Prevention Control
ISO	International Organisation for Standardisation
ISWA	International Solid Waste Association
ITOPF	International Tank Owners' Pollution Federation
IUCN	International Union for the Conservation of Nature
IUPAC	International Union of Pure and Applied Chemistry
IWM	Institute of Wastes Management
LAAPC	Local Authority Air Pollution Control
LAWDC	Local Authority Waste Disposal Company
LCA	Life Cycle Analysis
LIFE	L'Instrument Financiel pour l'Environnement (the EC Financial Instrument for Environment Established by Regulation 1973/93)
LNG	Liquefied Natural Gas
LNR	Local Nature Reserve

LPA	Local Planning Authority
LPG	Liquefied Petroleum Gas
LUST	Leaking Underground Storage Tank
LWRA	London Waste Regulation Authority
MAFF	Ministry of Agriculture, Fisheries and Food
MARPOL	International Convention for the Prevention of Pollution from Ships
MEL	Maximum Exposure Limit
MNR	Marine Nature Reserve
MPG	Mineral Policy Guidance
MSW	Municipal Solid Waste
MTBE	Methyl Tertiary Butyl Ether
NACCB	National Accreditation Council for Certification Bodies
NAWDC	National Association of Waste Disposal Contractors
NAWR	Noise at Work Regulations
NCC	Nature Conservancy Council
NERC	Natural Environment Research Council
NFFO	Non-Fossil Fuel Obligation
NGO	Non-Governmental Organisation
NIHHS	Notification of Installations Handling Hazardous Substances Regulations 1982
NNR	National Nature Reserve
NOEL	No Observed Effect Level
NRA	National Rivers Authority
NRPB	National Radiological Protection Board
NSCA	National Society for Clean Air and Environmental Protection
OECD	Organisation for Economic Co–operation and Development
OEL	Occupational Exposure Limit
OES	Occupational Exposure Standard
OFFER	Office of the Director General of Electricity Supply
OFGAS	Office of Gas Services
OFWAT	Office of Water Services
OJ	Official Journal of the European Communities
PAH	Polycyclic Aromatic Hydrocarbons
PAN	Planning Advice Note

PCB	Polychlorinated Biphenyls
PCH	Polychlorinated Hydrocarbons
PCP	Pentachlorophenol
PCT	Polychlorinated Terphenyl
PET	Polyethylene Terephthalate
PFA	Pulverised Fuel Ash
PIC	Prior Informed Consent
PPE	Personal Protective Equipment
PPG	Planning Policy Guidance
PVC	Polyvinyl Chloride
RCEP	Royal Commission on Environmental Pollution
RCRA (US)	Resource Conservation and Recovery Act 1976
RDF	Refuse Derived Fuel
RID	Regulations Concerning the International Carriage of Dangerous Goods by Rail
RIMNET	Radioactive Incident Monitoring Network
RIVPACS	River Invertebrate Prediction and Classification System
RMA (NZ)	Resource Management Act 1991
RSA	Radioactive Substances Act 1993
RSPB	Royal Society for the Protection of Birds
RTRA	Road Traffic Regulation Act 1984
RWMAC	Radioactive Waste Management Advisory Committee
SAC	Special Area of Conservation
SARA (US)	Superfund Amendments and Reauthorisation Act 1986
SI	Statutory Instrument
SEPA	Scottish Environmental Protection Agency
SNH (Scotland)	Scottish Natural Heritage
SOAFD	Scottish Office Agriculture and Fisheries Department
SOEmD	Scottish Office Environment Department
SPA	Special Protection Area
SRO	Scottish Renewables Order
SSSI	Site of Special Scientific Interest
SWQO	Statutory Water Quality Objective
TEL	Tetraethyl Lead
TBT	Tributyltin Compound
TCE	Tricochloroethylene
TCPA	Town and Country Planning Act 1990

TEM	Toluene Extractable Matter
TLV	Threshold Limit Value
TOC	Total Organic Carbon
TOVALOP	Tanker Owners' Voluntary Agreement Concerning Liability for Oil Pollution
TPH	Total Petroleum Hydrocarbons
TPO	Tree Preservation Order
TRI	Toxic Release Inventory
UCO	Town and Country Planning (Use Classes) Order 1987
UKELA	United Kingdom Environmental Law Association
UNCED	United Nations Conference on Environment and Development
UNECE	United Nations Economic Commission for Europe
UNEP	United Nations Environment Programme
UST (US)	Underground Storage Tank
UV	Ultraviolet
UWWTD	Urban Waste Water Treatment Directive
VOCs	Volatile Organic Compounds
WAMITAB	Waste Management Industry Training Advisory Board
WCA	Waste Collection Authority
WDA	Waste Disposal Authority
WHO	World Health Organisation
WIA	Water Industry Act 1991
WML	Waste Management Licence
WMP	Waste Management Paper
WO	Welsh Office
WQO	Water Quality Objective
WRA	Waste Regulation Authority
WRA 1991	Water Resources Act 1991
WWF	World Wide Fund for Nature

THE CONCISE LEXICON OF ENVIRONMENTAL TERMS

The European Union (EU) has been referred to throughout the *Lexicon* entries, for the sake of consistency and clarity for the reader, as the European Communities (EC). The reason for this is that European Union only became the accepted term for the European Communities after the signing of the Maastricht Treaty in November 1993.

abandoned mine

See *mine, abandoned.*

abatement

The control, reduction and/or removal of those emissions to the environment which have the potential to cause pollution.

abatement notice

Notice served by a local authority under the Environmental Protection Act 1990, s.80 requiring the abatement *(qv)* of a statutory nuisance *(qv)*, or prohibiting or restricting its occurrence or recurrence, and requiring the undertaking of any works or other steps necessary for those purposes. The recipient of the notice has 21 days in which to appeal to the Magistrates' Court. Failure (without reasonable excuse) to comply with an abatement notice is an offence.

abatement order

An order which may be made by a Magistrates' Court under the Environmental Protection Act 1990, s.82, on a complaint made by a person aggrieved by a statutory nuisance *(qv)*, requiring its abatement. Prior notice, normally of not less than 21 days, must be given to the proposed defendant of the intention to make such a complaint. The process is similar to that in the case of an abatement notice, except that it is available where a local authority is unwilling to act, or where the complaint is against a local authority. See also *nuisance; statutory nuisance.*

absorption

A process in which one substance, usually a liquid or a gas, is taken into the body of another.

abstraction

In relation to water, the removal of water from a source of supply, such as surface waters (rivers, lakes, ponds) or groundwater *(qv)*. Controls over abstraction of water are imposed by common law and by statute. At common law, the entitlement of a landowner to abstract water from a shared resource draws a distinction between water flowing in a defined channel, and percolating groundwater. Broadly speaking, a riparian rights *(qv)* owner is entitled to receive an unimpeded flow of water in its natural course, but this is subject to the rights of other riparian owners to the reasonable enjoyment of it.

Water may be abstracted from *surface water*, such as a running stream, for ordinary domestic purposes and for watering cattle. If this exhausts the flow, the lower owner has no remedy. If, however, the use is for other purposes, such as irrigation, then there is an obligation to return the water to the stream substantially undiminished in quantity or quality. This effectively rules out spray irrigation. However, rights of lower owners to object to abstraction beyond these limits may have been lost through express agreement (ie, the lower owner may be bought out) or prescription (under which the failure of the lower owner(s) to object over a sufficiently long period will be taken by the law to amount to deemed consent by them).

An easement to abstract water may arise by prescription if the right has been enjoyed *de facto* for more than 20 years without significant interruption. Rights to abstract *groundwater* depend upon whether the water is flowing through a defined channel or percolating through the ground. If it is in a defined channel, then common law regards it as if it were in a stream above ground, so that abstraction rights are attenuated by the rights of downstream riparian owners. If the water is percolating, then it may be abstracted by an occupier regardless of the effect on other owners. Today these common law rights are subject to further restrictions imposed by statute: see *abstraction of groundwater; abstraction of surface water.*

abstraction of groundwater

The removal of water from underground strata, known as an aquifer *(qv)*. With a few exceptions, it is an offence under the Water Resources Act 1991, s.24 to abstract groundwater without an abstraction licence. Administration of this regime is a function of the National Rivers Authority *(qv)*, whose policy is to authorise abstraction only where:

1 total abstraction from any groundwater resource area does not exceed the long-term annual average rate of replenishment;

2 there is no unacceptable detriment to any watercourse or other environmental feature dependent upon groundwater; and

3 any abstraction does not cause a deterioration of groundwater quality through the incursion of saline or polluted waters (NRA, *Policy and Practice for the Protection of Groundwater*

(1992), p 25). There is no statutory licensing system for Scotland.

The total number of groundwater abstractions in England and Wales is estimated to be in excess of 100,000. There are nearly 2,000 major public supply sources, and abstracted groundwater accounts for around 35% of public water supply in England and Wales. 'Abstraction' is broadly defined by s.221(1), in relation to water contained in any source of supply, as 'the doing of anything whereby any of that water is removed from that source of supply, whether temporarily or permanently, including anything whereby the water is so removed for the purpose of being transferred to another source of supply'. There can be only one point of abstraction, and the fact that by abstracting from a particular point a licence holder may draw water from other parts of a hydrologically linked system does not impose any duty to obtain a further abstraction licence in respect of other linked sources of supply: *British Waterways Board v National Rivers Authority* [1993] Env LR 239 (CA).

abstraction of surface water

The removal of water from a surface source of supply as opposed to groundwater. The common law rights of riparian owners to abstract water are subject to a licensing regime under the Water Resources Act 1991, which is the responsibility of the National Rivers Authority. There is no parallel system in Scotland. However, the Act underwrites common law rights of abstraction from inland waters by or on behalf of the owner of contiguous land for use on that land (with or without other land held with that land) for the domestic purposes of the occupier's household or for agricultural purposes (other than spray irrigation) and not exceeding 20 cubic metres in any 24 hour period (s.27(3) and (4) as applied by s.27(6) and s.39(3)). This is a 'protected right', together with a right to abstract in accordance with an abstraction licence, and the Authority is enjoined not to grant further abstraction licences that will interfere with them, on pain of paying damages for breach of statutory duty. See also *abstraction; abstraction of groundwater; riparian rights.*

acceptable daily intake (ADI)

In the context of pesticides, the amount of a chemical which can be consumed every day for an individual's entire lifetime, in the practical certainty, on the basis of all known facts, that no harm will result. The ADI is expressed as milligrams of chemical per kilogram of consumer body weight. ADIs for pesticides are set by the Joint Food and Agriculture Organisation/World Health Organisation Meeting on Pesticide Residues.

Accord Européen Relatif au Transport International des Marchandises Dangereuses par Route (European Agreement on the International Carriage of Dangerous Goods by Road) (ADR)

A treaty regulating the transportation of dangerous substances and articles by road in Europe. It sets out requirements relating to the packaging and labelling of goods and the vehicles in which they are transported.

accredited environmental verifier

An organisation or individual authorised to validate the environmental policies, programmes, management systems, reviews, audit procedures and statements prepared by a business which is complying with the EC Eco-management and Auditing Scheme (EMAS) *(qv)*. Such person must be independent of the business being verified, and must satisfy the requirements and carry out the functions set out in Annex III to the Regulation. Member States were required to establish systems for the accreditation and supervision of independent environmental verifiers by April 1995. The National Accreditation Council for Certification Bodies *(qv)* has responsibility for the establishment of this system in the United Kingdom.

acid rain

The deposition (wet, dry or occult, the latter meaning via mist or fog) of acidic compounds in the atmosphere. The main sources of the compounds are the sulphur dioxide and oxides of nitrogen produced by the burning of fossil fuels in power stations, oil refineries or combustion plants, and from motor vehicles.

activated carbon

A form of carbon with a porous, honeycombed structure and a very high internal surface area. Activated carbon has a high absorptivity for gases, vapours and colloidal solids and is used widely in pollution control, eg in air and water purification, waste treatment, solvent recovery, the removal of sulphur dioxide from stack gases and air conditioning. It is flammable and toxic to humans as a dust inhalation hazard.

activated sludge

Flocculent (the coagulation of small particles into larger particles) sludge produced by the growth of bacteria *(qv)* and other organisms in raw or settled sewage *(qv)* when it is continuously aerated *(qv)*. See also *aeration; sewage sludge*

activated sludge process

A biological sewage treatment process in which a mixture of sewage and activated sludge *(qv)* is activated and aerated. The activated sludge is subsequently separated from the treated sewage by sedimentation and may be reused. See also *aeration; sewage.*

active ingredient

The component of a product, the activity of which makes that product fit for its intended use.

actus reus

One of the two elements whose existence must normally be proved to constitute a crime, which involves a mental element or *mens rea*, and an act or omission (the *actus reus*). Crimes in which *mens rea* is not a necessary element are sometimes known as offences of strict liability *(qv).*

ADR

See:

1 *Accord Européen Relatif au Transport International des Marchandises Dangereuses par Route;*

2 *alternative dispute resolution.*

adsorption

Process in which materials adhere to the surface of a solid body, usually in layers a few molecules thick *(cf* absorption *(qv)*). In relation to sewage *(qv)* and effluent *(qv)* treatment, the process is usually employed for the treatment of solutions to remove dissolved materials.

advanced transport technology

Technological approaches to reducing travel need and time by improvements to public transport, telecommuting, teleworking, teleshopping, providing up-to-date traffic information and minimising transport effects by automatically diagnosing and controlling emissions.

Advisory Committee on Business and the Environment (ACBE)

Committee established by the Department of the Environment (DOE) and the Department of Trade and Industry (DTI) in 1991 for two years (then re-established for a further two-year term) comprising representatives from all sectors of the business community. The aims of the ACBE are to provide a forum for the discussion of environmental issues between Government and business, to promote good environmental practice and management in the business community and to highlight international business initiatives on the environment.

Advisory Committee on Hazardous Substances (ACHS)

Committee established by the Secretary of State under the Environmental Protection Act 1990, s.140(5) and Sched 12 to give him advice on making regulations as to the importation, use, supply or storage of any substances with potential to cause harm to the environment or to human health (under s.140) and on the obtaining of information on potentially hazardous substances under s.142). The Committee comprises representatives of industry, consumer and environmental protection groups. The Committee also advises on the risk assessment of chemical substances.

Advisory Committee on Releases to the Environment (ACRE)

Committee appointed by the Secretary of State for the Environment under the Environmental Protection Act 1990, s.124 to advise on matters relating to the regulation of genetically modified organisms (GMOs) *(qv)* under Part VI of the Act. In particular, the Committee advises on applications for consent to release or market GMOs and on the conditions which should be attached to such consents if granted.

Advisory Group on Eco-management and Audit (AGEMA)

Committee appointed to advise the Secretary of State for the Environment in his capacity as a Competent Body for the purposes of the EC Eco-management and Auditing Scheme (EMAS) *(qv)*. AGEMA's terms of reference are to consider the operational aspects of the EMAS scheme, including the interrelationship between the Competent Body, the National Accreditation Council for the Certification Bodies and the enforcement authorities.

aeration

The bringing about of intimate contact between air *(qv)* and liquid by one of several methods, for example spraying the liquid into the air or forcing air through the liquid, or agitating the liquid to promote surface absorption *(qv)* of air.

aerobic

Operating in the presence of oxygen; requiring oxygen for respiration. The opposite of anaerobic *(qv)*.

aerosol

A particle of solid or liquid matter of such small size that it can remain suspended in the atmosphere for a long period of time. Aerosols diffuse light and the larger particles settle out on horizontal surfaces or cling to vertical surfaces.

afforestation

The planting of land with trees for forestry purposes. Afforestation does not normally require planning permission *(qv)*, but certain projects require the preparation of an environmental statement *(qv)* under the Environmental Assessment (Afforestation) Regulations 1988 (SI 1988 No 1207) (in Scotland, the Environmental Assessment (Scotland) Regulations 1988 (SI 1988 No 1221)).

Agenda 21

A worldwide programme of action for sustainable development *(qv)* into the 21st century which was adopted by the governments attending the United Nations Conference on Environment and Development (UNCED) *(qv)* in Rio de Janeiro in June 1992. Agenda 21 requires action by governments, the United Nations, non-governmental organisations and other bodies and individuals. Agenda 21 addresses the social, economic and legal dimensions of sustainable development, the conservation and management of the world's resources, the roles of the different groups involved and the implementation of this programme. See also *Local Agenda 21*.

Agricultural Development and Advisory Service (ADAS)

A Next Steps (ie Government initiated and sponsored) executive agency which in England and Wales provides advisory, consultancy and information services to the Government, public authorities, agricultural bodies and agriculture-related industries. One ADAS objective is to control water pollution arising from farm wastes. Farmers are often assisted by ADAS in the preparation of their own farm waste management plans, which address the issues of spreading manure and comparable organic wastes on land in the most economical and environmentally friendly manner.

agricultural pollution

Pollution originating from farms or other agricultural enterprises. Air pollution from agricultural sources may include odours, smoke pollution, the production of greenhouse gases and the emission of ammonia from manure and slurry. Pollution of watercourses may result from the discharge of sewage, slurry and run-off of phosphates or nitrates as a result of the use of agrochemicals *(qv)*. Poor agricultural practices have deleterious effects on soil fertility and lead to soil erosion, and pollution is also caused by agricultural waste and noise. These problems are addressed by EC Regulation 2078/92 on agricultural production methods compatible with the requirements of the protection of the environment and the maintenance of the countryside.

agri-environmental regulation

EC Regulation 2078/92 which formed part of the 1992 reforms of the Common Agricultural Policy, and which has the objectives of reducing the polluting effects of agriculture, encouraging environmentally favourable extensification of crops and livestock and protecting the countryside, recovering abandoned land, setting up environmentally beneficial long-term set-aside arrangements, facilitating public access to agricultural land and educating farmers and landowners on agri-environmental matters. The regulation specifies a number of measures which may bring about these objectives, including changes in farming practices. Member States are required to draw up five-year zonal programmes, containing detailed environmental and agricultural information about each zone. Further reference: House of Lords Select Committee on the European Communities, *Environmental Aspects of the Common Agricultural Policy* (Session 1992–93; HL Paper 45).

agrochemicals

Chemical products used in connection with agricultural operations, eg farms or other agricultural enterprises. This generic term includes such products as fertilisers, pesticides, herbicides, insecticides and plant growth regulators. Waste agrochemicals derived from a farm are not classified as controlled waste *(qv)* under the Environmental Protection Act 1990, s.75, and agricultural waste generally is not presently directive waste *(qv)*.

air

The mixture of gases, mainly nitrogen (N_2) and oxygen (O_2), forming the Earth's atmosphere. One of the three environmental media for the purposes of the Environmental Protection Act 1990, where it is defined for certain purposes of the Act so as to include air within buildings and other natural or man-made structures, whether above or below ground.

aircraft noise

Literally, the noise made by aircraft on the ground or overflying; in practice, the noise perceived by humans as emanating from aircraft. Measured in a variety of different decibel units, depending on purpose; eg EPNdB or effective perceived noise is internationally used for aircraft classification, accounting for aircraft approaching, being overhead, and going away; L_{eq} as an equivalent continuous sound level over a specified measurement period.

Noise certification requirements are internationally agreed; see EC Directive 92/14, implemented in the UK by the Aeroplane Noise

(Limitation on Operation of Aeroplane) Regulations 1993 (SI 1993 No 1409). Trespass and nuisance actions against civilian aircraft in normal flight are precluded by the Civil Aviation Act 1982, s.76. Orders are made under s.78 to regulate noise and vibration and alleviate its effects, including restrictions on night flying. For discussion of these restrictions, see *R v Secretary of State for Transport ex parte London Borough of Richmond-upon-Thames* [1994] 1 All ER 577.

air pollution

Air pollution can be defined as the presence in the external atmosphere of one or more contaminants (pollutants) *(qv)*, or combinations thereof, in such quantities and of such duration as may have the potential to be injurious to human health, plant or animal life, or property (materials); or which unreasonably interferes with the comfortable enjoyment of life, property, or the conduct of business.

The main air pollutants in urban Britain have changed significantly over the past half-century, since the cleaning up of the emitters of black smoke and sulphur dioxide produced by industrial and domestic coal burning under the Clean Air Acts *(qv)*. There has been a rapid rise since the early 1980s in emission of oxides of nitrogen, hydrocarbons, carbon monoxide, respirable particulates, volatile organic compounds (VOC) *(qv)* and the secondary pollutant, ozone. Further reference: House of Commons Transport Committee, *Transport-related Air Pollution in London* (Session 1993–94; HC 506). See also *atmospheric pollution*.

air pollution control

See *integrated pollution control; local authority air pollution control*.

air quality management area

An area to be declared by a local authority, under proposals advanced by the Government in 1995, where it has been shown that the Government's targets for air quality are not likely to be met solely by national policies, and for which the local authority will be required to draw up air quality management plans setting out their proposals to meet the targets. They will also have a duty to appraise their development, transport and pollution control policies against the air quality assessment in the designated areas. Further reference: DOE Consultation Paper, *Air Quality: Meeting the Challenge* (1995).

air quality standard

An environmental quality standard *(qv)* in relation to air, normally a quantitative standard which prescribes maximum and/or permissible concentrations of substances, such as carbon monoxide, fine particulates, nitrogen oxide and sulphur dioxide. Statutory air quality standards are in place in the United Kingdom under the Air Quality Standards Regulations 1989 (SI 1989 No 317), which implement EC Directives 80/779, 82/884 and 85/203, and which require the Secretary of State to take any appropriate measures to ensure that concentrations of sulphur dioxide, nitrogen dioxide and suspended particulates in the atmosphere do not exceed the limit values *(qv)* prescribed in the Directives, and set a maximum mean annual value for the concentration of lead in the air. The Regulations also oblige the Secretary of State to maintain measuring stations to supply the data necessary for the Directives.

aldrin

A white insecticide containing a chlorinated derivative of naphthalene, $C_{12}H_8Cl_6$. It is especially effective against pests resistant to dichloro diphenyl trichloroethane (DDT) *(qv)*. See also *red list*.

algal bloom

A phenomenon occurring in water under certain conditions as a result of the abundant growth of phytoplankton or zooplankton where nutrient enrichment of the waters due to the addition of nitrogen or phosphorus (eg via run-off) has led to eutrophication (qv) and an unnatural imbalance in animal and plant diversity. Blooms can restrict the entry of light and cause oxygen depletion at night, killing the other organisms present. They may also form scums around water course margins which are unsightly and, in some circumstances, toxic. The quality of water for abstraction, conservation and amenity can all be severely affected by algal bloom.

alkali

A substance which upon dissolving in water produces more hydroxyl than hydrogen ions and neutralises acids to form salts.

alkali works

Industrial works which were controlled under early British environmental legislation, the Alkali Act 1863, the Alkali Etc. Works Regulation Act 1906 and much later also by the Health and Safety at Work Etc. Act 1974. Alkali works, mostly producing sodium carbonate from salt, were commercially viable from the 1820s, but the process produced large volumes of hydrogen chloride gas and an unpleasant smell.

The problem initially arose in the Sefton area of Merseyside. In 1836, a packed tower was patented (the ancestor of scrubbers *(qv)*) but these towers still allowed up to half the gases to escape. Until 1862,

Parliament and local authorities had attempted to deal with such nuisances simply by banning them, without making any constructive suggestions on how to abate them. The 1862 Royal Commission and the 1863 Alkali Act, which implemented the Commission's recommendations, took a new approach. The Royal Commission recommended that if at least 95% of the hydrochloric acid gas evolved from alkali works were arrested, the remainder, after adequate dilution, could be allowed to pass into the air. The Alkali Act 1863 required that works within its ambit should use the best practicable means *(qv)* to reduce to the minimum the discharge of noxious or offensive gases, and required all such works to be registered.

This system is being progressively superseded over the period 1991–96 by the integrated pollution control (IPC) *(qv)* and local authority air pollution control (LAAPC) *(qv)* systems introduced under Part I of the Environmental Protection Act 1990.

alternative dispute resolution (ADR)
Informal arrangements outside traditional legal forms for resolving disputes (typically, in the environmental sphere, those arising between industrial or development interests and local communities), including negotiation, conciliation, consultation and arbitration, with a view to avoiding the potentially confrontational, time-consuming and expensive forms of formal dispute resolution through the courts.

ambient air quality
The quality of the surrounding air *(qv)* in any particular environment.

Amoco Cadiz
The name of a supertanker which grounded off northern Brittany on March 16, 1978, discharging 223,000 tonnes of Arabian crude oil, causing the death of thousands of birds and extensive damage to marine life and the coast.

anaerobic
Operating in the absence of oxygen; able to respire without oxygen. The opposite of aerobic *(qv)*.

anthropogenic
Resulting from or influenced by human activity or intervention.

appeal
An application to a higher body in an administrative or judicial hierarchy to reconsider a decision taken lower in the hierarchy. In

environmental legislation, applicants for permits normally have a right to appeal from decisions of regulators to central Government (represented by the appropriate Secretary of State): eg a planning appeal against a decision of a local planning authority *(qv)*, or against a decision by the National Rivers Authority relating to a discharge consent *(qv)* or an abstraction *(qv)* permit; or against a decision by Her Majesty's Inspectorate of Pollution relating to an integrated pollution control (IPC) authorisation. On such an appeal, all the merits of the application are reviewed again, and the appellant or the regulator is entitled to insist upon the holding of a public local inquiry, although it is common for appeals to be dealt with instead by written representations. There is also normally a further right to apply or appeal to the High Court, but limited to a point of law (such as a failure to comply with proper procedures, or to take all relevant factors into account).

Approved Code of Practice (ACOP)
A term applied to a code of practice *(qv)* approved by the Health and Safety Commission with the consent of the Secretary of State for Employment, under the Health and Safety at Work Etc. Act 1974, s.16 for the purpose of providing practical guidance on compliance with legislation. Although the failure to observe the provisions of an ACOP does not of itself give rise to criminal liability *(qv)* nor to civil liability *(qv)*, such failure may normally be taken into account by a court in criminal proceedings as proof of contravention of the legislation to which the ACOP relates unless a person can satisfy the court that he complied with the legislation in some other way.

aquiclude
A geological formation of rock or soil which, although porous and capable of absorbing water slowly, will not transmit it fast enough to furnish an appreciable supply for a spring or a well.

aquifer
A permeable underground rock formation or stratum that is capable of storing and transmitting groundwater in significant quantities. Aquifers are effectively underground reservoirs. The containment of the water is because of the impermeable nature of the walls of the aquifer, which are known as aquicludes *(qv)*. The characteristics of an aquifer (eg whether it is fissured, and its porosity) will determine its vulnerability to pollution and whether it can be exploited for water supply. See also *abstraction of groundwater; groundwater.*

aquifer protection zone
See *source protection zone.*

aquitard

A geological formation of rock or soil which retards the movement of groundwater.

area of archaeological importance (AAI)

An area designated under the Ancient Monuments and Archaeological Areas Act 1979, Part III, as being of archaeological importance. It is intended to be used in areas, like the centres of ancient towns, where it is known or suspected that there is extensive buried archaeology but where the remains are not sufficiently discrete to qualify for protection as a scheduled monument. Designation does not of itself protect the area from damaging development, but it allows the local planning authority to insist upon a delay in the commencement of operations for a sufficient time to allow rescue of the archaeology on the site. Six weeks notice must be given to the local authority before development *(qv)* in an area of archaeological importance can begin. An investigating authority, normally an archaeological unit, may examine, excavate and record the site for up to four months and two weeks from the expiry of the notice.

area of local wildlife importance

A non-statutory designation used in development plans *(qv)* to define an area (less important than a county site *(qv)*) to which protective policies are applied in the plan for nature conservation. Sometimes known as area of nature conservation importance.

area of outstanding natural beauty (AONB)

An area of countryside designated by the Countryside Commission (in Wales, the Countryside Council for Wales) and approved by the Secretary of State under the National Parks and Access to the Countryside Act 1949, s. 87. The main protection for AONBs is under the planning system. Planning policy guidance (PPG) *(qv)* notes contain general protective policies eg PPG7 *The Countryside and the Rural Economy* which contains policies against major development in AONBs. Designations are reflected in appropriate policies in development plans *(qv)*, and permitted development *(qv)* rights under the General Development Order *(qv)* are more tightly drawn. For Scotland, see also *national scenic area.*

article

The name given:

1 to particular paragraphs in orders made by Ministers under statutory authority (see also *regulations*);

2 to paragraphs in international treaties, including the Treaty of Rome, and EC Directives *(qv)*.

Article 4 direction

An order made under Article 4 of the General Development Order *(qv)* removing specified permitted development rights in relation to a specified area (eg rights to extend dwelling houses in a conservation area).

asbestos

The collective name given to a group of naturally occurring impure mineral fibres (of magnesium silicate). Asbestos is chemically inert, heat resistant and mechanically strong and therefore has a wide range of uses. Types of asbestos are: anthophyllite, tremolite, actinolite, amosite and crocidolite (these are all known as amphibole asbestos); and crysotile (serpentine asbestos). Crocidolite (blue asbestos), amosite (brown asbestos) and crysotile (white asbestos) are of most commercial significance and therefore the most commonly used. The entry of asbestos fibres into the body is related to a variety of diseases, particularly of the respiratory and gastro-intestinal tracts. A range of legislation covers the use, supply, marketing and disposal of asbestos, principally the Control of Asbestos at Work Regulations 1987 (SI 1987 No 2115).

Deaths related to asbestos in the UK will not peak for at least 20 years according to a joint study team reporting in 1995 at the Institute of Cancer Research. The Health and Safety Executive in March 1995 has also warned that the death toll from lung cancer and mesothelioma, which is cancer of the lining of the chest or stomach, could rise from 5,000 a year to between 5,000 and 10,000 deaths a year by 2020. It has been held judicially that a contractor undertaking work on materials containing asbestos should have been aware at the latest by 1988 of the associated health hazard: see *Barclays Bank plc v Fairclough Building (No 2). The Times,* February 15, 1995.

The pattern of the asbestos cancer epidemic has been quite different in the US where use of the materials reached a peak in the 1940s. US death rates in 1995 are already beginning to fall.

The main victims will be people who have worked in the building industry and associated trades in the 1950s, 1960s and 1970s when asbestos was a popular material for construction and insulation. Substantial exposure may still be occurring in the building, renovation and maintenance trades.

Association for the Protection of Rural Scotland

A non-governmental organisation which has similar aims and functions in respect of the Scottish rural environment as the Council for the Protection of Rural England *(qv)*.

atmospheric pollution

Air pollution at local, regional, national or global level. Attempts to control air pollution were originally directed at alleviating the local effects of smoke emission, but today's approach must address also areas of regional (eg acid rain *(qv)*) and global (eg greenhouse gases *(qv)*) concern. Pollution of the air by smoke has been a problem ever since mankind first began using fire for heating, cooking and metalworking. In 1306 a Royal Proclamation prohibited London's artificers from using seacoal in their furnaces; one offender was executed.

The principal controls over emissions today are those restricting the emission of smoke from industrial and domestic sources (Clean Air Act 1993), emission of substances from certain scheduled processes (integrated pollution control and local authority air pollution control under the Environmental Protection Act 1990), and controls over products, such as motor vehicle design and operation.

attenuation

Reduction, dilution; eg, the breakdown or dilution of a contaminant in water.

bacteria

Primitive vegetable organisms which reproduce by division, and are found in decomposing animal and vegetable liquids. There are three principal types:

1 aerobic *(qv)* bacteria which require free oxygen for their growth;

2 anaerobic *(qv)* bacteria which grow in the absence of free oxygen and which derive oxygen by breaking down complex substances; and

3 coliforms *(qv)*, a type of bacteria found in sewage and decaying vegetation.

Basle (or Basel) Convention

A global convention, which entered into force on March 22, 1989 in Basle, on the transboundary movements of hazardous wastes and their disposal. The Convention aims to help protect the environment through more significant controls on the transboundary movement of hazardous waste and its management. It recognises the right to prohibit the import of hazardous waste. Where the import of waste has been prohibited, parties may not export such waste unless there is written consent to the particular transportation. Any transboundary movement of waste must be accompanied by sufficient details for its environmental impact to be assessed. The European Communities (EC) is a signatory to the Convention and adopted its principles in EC Regulation (259/93) on the supervision and control of waste within, into and out of the European Communities. Amongst other conventions arising out of the Basle Convention is the 1991 Bamako Convention on the control of the transboundary movement of hazardous wastes within Africa. Adopted by Member States of the Organisation of African Unity , it bans the import of hazardous waste into the signatory African states from countries outside Africa, and strictly controls the movements of waste between those states.

bathing waters

All waters (inland or coastal) except those intended for therapeutic purposes or used in swimming pools, and areas either in which bathing is explicitly authorised by Member State authorities, or in which bathing is not prohibited and is traditionally practised by a large number of bathers (EC Directive 76/160 concerning the quality of bathing water). In the United Kingdom, bathing water is defined according to the 'traditional areas' description and the Secretary of State for the Environment has now designated over 400 bathing areas (having originally been able to find only 27, and excluding Blackpool and Brighton). Water in such areas must meet specified quality standards relating to chemical, microbiological and physical parameters.

The Directive is currently implemented in England and Wales through the granting of consents under the Environmental Protection Act 1990, and the Water Resources Act 1991, and in accordance with the Bathing Waters (Classification) Regulations 1991 (SI 1991 No 1597) which prescribe water quality objectives *(qv)* for bathing waters. A proposal for a directive to modify and streamline the Directive was published by the European Commission in 1994 for consultation, and is the subject of a special report by the Select Committee on the European Communities of the House of Lords, *Bathing Water* (Session 1994–95; First Report; HL Paper 6). See also *coliform*. Further reference: National Rivers Authority, *Bathing Water Quality in England and Wales – 1993* (1994).

becquerel (Bq)

A unit of radionuclide activity equivalent to one disintegration per second (dps). This term is usually used instead of the curie which is equivalent to 3·7 x 10^{10} dps. It does not reflect the harmfulness of the type of radiation emitted as a result of the disintegration, simply the rate of disintegration. In 2,100 homes in the UK the average concentration of radon was measured as 20 Bq per m^3, with most below 100 Bq per m^3 but including 48 above this level. The National Radiological Protection Board advise that remedial action be taken when levels exceed 400 Bq per m^3. This degree of exposure corresponds to a 2% risk of death from lung cancer in non-smokers. Further reference: House of Commons Environment Committee, *Indoor Pollution* (6th Report, Session 1990–91; HC 61).

benthos

Animals and plants living on the bottom of a sea or lake, attached or unattached, from the deepest levels up to the high water mark. Benthic organisms are divided into littoral and deep water organisms.

bentonite

A clay mineral in a group known as smectites. They are referred to as high-swelling clays because their mineral structure expands hugely when they absorb water. The properties of bentonites vary depending upon the cations that are associated with them, such as sodium or calcium. Sodium bentonite is the most commonly used for landfill applications.

benzene

A colourless liquid hydrocarbon obtained from coal and petroleum. It is often used in the manufacture of organic chemicals, as a solvent and as a fuel. The presence of benzene in the atmosphere, such as from the combustion of unleaded fuel without a catalytic converter *(qv)* is a hazard to human health.

Berne Convention

The Convention on the conservation of European wildlife and natural habitats which was agreed in Berne on September 9, 1979 and entered into force on June 1, 1982, to which the European Communities (EC) is a signatory. The aims of the Convention are to conserve wild flora and fauna and their natural habitats, to promote co-operation between countries in their conservation efforts and to protect endangered species. EC Directive 92/43 (the Habitats Directive *(qv)*) on the conservation of natural habitats and of wild fauna implements many of the principles of the Convention.

best available techniques not entailing excessive cost (BATNEEC)

A statutory formula used in controlling emissions to the

environment which seeks a balance between requiring use of state of the art technology and processes to minimise emissions, and the cost of doing so. It is an implied general condition of every authorisation granted under the Environmental Protection Act 1990, Part I on integrated pollution control (IPC) *(qv)* and air pollution control *(qv)* that in carrying on a prescribed process, BATNEEC will be used for preventing or reducing the release of substances harmful to the environment or rendering harmless any other substances released. Guidance on what BATNEEC means in practice is given in Part 7 *Integrated Pollution Control. A Practical Guide* (HMSO, 1993). See also *prescribed processes and prescribed substances.*

best available technology (BAT)

A statutory formula used in controlling emissions to the environment which requires the use of the best technology available to prevent or minimise emissions, irrespective of its cost. It was introduced under the Federal Clean Water Act which required new services to comply with BAT from the outset, and existing sources to do so by 1983; and in Germany in 1986 under water legislation, superseding the former test of 'generally accepted technological standards'. The formula is often restricted to technology which would not entail excessive cost (see eg EC Directive 84/360 on air pollution from industrial plants and, in the Environmental Protection Act 1990, it is adapted to a requirement to use the best available techniques not entailing excessive cost BATNEEC) *(qv)* though for new plant and processes Her Majesty's Inspectorate of Pollution (HMIP) insists generally upon BAT.

best environmental option (BEO)

The option which, in the context of releases to the environment, provides the greatest benefit or least damage to the environment as a whole, irrespective of cost, in the long term as well as the short term. Its importance lies in its holistic focus, rather than upon attempting to regulate separately emissions to different environmental media. Applied in the United Kingdom in a qualified form as best *practicable* environmental option (BPEO) *(qv)*, which requires regulators to have regard also to considerations of cost.

best practicable environmental option (BPEO)

The option that provides the best benefit or least damage to the environment as a whole at acceptable cost, in the long term as well as the short term. BPEO was first developed as a concept in the 5th Report of the Royal Commission on Environmental Pollution (RCEP) *(qv)* and can be regarded as a development of the best environmental option *(qv)* concept, which links it to economic objectives and takes a long-term approach to environmental solutions. The Environmental Protection Act 1990 (s.7(7)) requires that BPEO be used for minimising pollution of the environment as a whole as respects substances released in a process subject to integrated pollution control *(qv)*. Further reference: *Best Practicable Environmental Option – A New Jerusalem?* (United Kingdom Environmental Law Association (UKELA)); Royal Commission on Environmental Pollution 12th Report, *Best Practicable Environmental Option* (Cm 310; 1988).

best practicable means (BPM)
A statutory formula used in controlling emissions to the environment and which requires the use of the best means available to secure a particular objective. First introduced by the Alkali Act 1863, it came to symbolise the pragmatic British approach to balancing investment in environmental improvement against considerations of cost. Although soon to be superseded by the Environmental Protection Act 1990 (EPA 1990), BPM has also played a key part in air pollution control under the Health and Safety at Work Etc. Act 1974 (s.5). BPM is retained under s.80(7) of the 1990 Act as a defence in certain proceedings relating to statutory nuisances, if the defendant can show that the best practicable means were used to prevent or counteract the effect of the nuisance. Some guidance on its meaning in this context is found in s.79(9) EPA 1990 which defines 'practicable' as reasonably practicable having regard among other things to: local conditions and circumstances; the current state of technical knowledge; and the financial implications. 'Means' includes design, maintenance, installation and operation of plant and machinery, and the design, construction and maintenance of buildings and structures.

best technical knowledge not entailing excessive costs (BTKNEEC)
A principle to be applied, under EC Directive 91/271 on urban waste water, to the design, construction and maintenance of urban waste water collecting systems. Factors to be taken into account include the volume and characteristics of urban waste water, the prevention of leaks, and limiting pollution to receiving waters due to sewer overflow.

Bhopal
A town in India where, on December 3, 1984, a storage tank exploded in a chemicals factory owned by Union Carbide and released a cloud of methyl isocynate over the town, causing over one thousand deaths and as many as 250,000 injuries.

bio-accumulatable
Having the ability to accumulate in the food chain *(qv)*.

biochemical (or biological) oxygen demand (BOD)
The amount of dissolved oxygen consumed by chemical and microbiological action when a sample is incubated for five days at 20°C. The BOD normally gives a rough indication of the organic matter present in the sample and is often used as an indicator and measure of the quality of water particularly in respect of trade effluent. See also *chemical oxygen demand (COD); microbiology.*

biocides
A diverse group of poisonous substances including preservatives, insecticides, disinfectants and pesticides used for the control of organisms that are harmful to human or animal health or that cause damage to natural or manufactured products. The European Commission has published a proposal for a directive regulating the placing of biocidal products on the market (COM (93) 351 final – SYN465).

biodegradable
Having the ability to decompose through the action of bacteria or other organisms.

biodiversity
The abbreviation of biological diversity. A definition is provided by Article 2 of the Biodiversity Convention: 'the variability among living organisms from all sources including, *inter alia,* terrestrial, marine and other aquatic ecosystems and the ecological complexes of which they are part. This includes: diversity within the species, between the species and of ecosystems.' *Biodiversity: The UK Action Plan* Cm 2428 (1994) explains the commitments made at the United Nations Conference on Environment and Development *(qv)* and the objectives of the Action Plan. The conservation of biodiversity has recently been recognised to be of great importance, both globally and in the United Kingdom. See also *Convention on Biological Diversity.*

biological oxygen demand (BOD)
See *biochemical oxygen demand.*

biomass
The total mass of living organisms in a given area or the organic material derived from or consisting of living organisms in a given area.

biosphere
The part of the Earth and its atmosphere in which living things are found.

biotechnology

The industrial application of biological systems or processes or organisms to solve problems and provide long-term sustainable solutions. The conversion of biological or other raw materials into products of greater value by the use of processes which involve organisms, tissues, cells, organelles or isolated enzymes.

Birds Directive

European Communities (EC) Directive 74/409 on the conservation of wild birds (as amended), the purpose of which is to establish a system of protection for all species of wild birds indigenous to Europe. The Directive also requires the provision of sufficient habitat diversity and area so as to be able to maintain the population of all species. See also *Habitats Directive; special protection area.*

black list

The substances specified in List I in the Annex to EC Directive 76/464 on pollution caused by certain dangerous substances discharged into the aquatic environment of the Communities. The obligation on Member States is to eliminate the pollution of waters by the groups and families of substances contained in the black list. See also *grey list; red list.*

Bonn Convention

The Convention on the conservation of migratory species of wild animals, agreed in Bonn on June 23, 1979.

borehole

A narrow shaft, normally drilled mechanically into the ground. Boreholes are widely used as sampling and monitoring points and are drilled into and occasionally below an aquifer to monitor groundwater *(qv)* quality. Boreholes for sampling may be one-off core sampling boreholes, or long-term monitoring points, (eg landfill gas-monitoring boreholes at the periphery of a landfill *(qv)* site). Boreholes may also be used for extraction of contaminants (eg landfill gas).

Braer

A Liberian-registered tanker which grounded on the Shetland Islands on January 5, 1993 and over the course of the following week discharged its cargo of 85,000 tonnes of Norwegian crude oil into the ocean. The severe weather conditions meant that the oil, which was of a type particularly suited to dispersion, did not form into a conventional oil slick, and there was almost no stranding of oil on the coastline. The major impact of the spill was on the marine environment rather than on land. Further reference: Scottish Office, *The Environmental Impact of the Wreck of the Braer* (1994).

British Government Panel on Sustainable Development

A panel appointed by the Prime Minister in January 1994 to advise him on key issues relating to sustainable development. The Government consults the Panel on issues of major importance, and the Panel has access to all Ministers. It published its first report in January 1995, concentrating on environmental pricing, environmental education, depletion of fish stocks and atmospheric ozone depletion. Parallel initiatives by the Prime Minister included the Round Table on sustainable development *(qv)* and the Citizens' Environment Initiative, now known as 'Going for Green' *(qv)*.

British Standard

An objective standard established by the British Standards Institution and promulgated with a view to avoiding duplication of designs, and promoting minimum standards. Two standards relevant to the environment are British Standard 7750 on a Specification for Environmental Management Systems (BS 7750) *(qv)* and British Standard 5750 on Quality Systems (BS 5750) *(qv)*. British Standards are widely recognised and are often used as the basis of European Communities (EC) and other international standards including those established by the International Organisation for Standardisation (ISO).

British Standard 5750 on Quality Systems

A standard established by the British Standards Institution against which the quality systems of an organisation can be assessed. BS 5750 is divided into three distinct parts: Part 1 Specification for design, manufacture and installation; Part 2 Specification for manufacture and installation; and Part 3 Specification for final inspection and test. Organisations that comply with the standard are registered by the British Standards Institution. BS 5750 is the basis upon which British Standard 7750 on a Specification for Environmental Management Systems (BS 7750) was developed.

British Standard 7750 on a Specification for Environmental Management Systems

A British Standard *(qv)* which specifies the requirements of an environmental management system *(qv)* against which a commercial operation can establish and assess a system to manage and improve its environmental performance. BS 7750 addresses the concept of an environmental management system as a whole and provides an organisation with a model management system against which it can develop and measure the capability of its own environmental management system. See also *Eco-management and Auditing Scheme (EMAS)*.

British thermal unit (Btu)

The amount of heat necessary to raise the temperature of one pound of water by 1° F (100,000 Btu = 1 therm). It equals 1055·06 joules.

Brundtland Report

Report of the 1987 World Commission on Environment and Development (chaired by Gro Harlem Brundtland) entitled *Our Common Future* (Oxford University Press, 1987). The Report adopted as its central theme the concept of sustainable development *(qv)* which is widely used as the basis for an integrated approach to both economic and environmental policies.

Building Research Establishment Environmental Assessment Method (BREEAM)

A procedure, developed by the Building Research Establishment, for assessing the design or construction of a building according to the extent to which environmental issues are addressed by that design/building. Separate assessment methods have been developed for each of new and existing offices, superstores and supermarkets, new industrial units and new homes. The assessment is separated into global effects, neighbourhood effects and indoor effects. Assessments are carried out in two parts; the first part relates to the building fabric and services while the second part relates to the operation and management of the building. Buildings which are assessed according to BREEAM are independently certified and rated.

cadmium

A metal found in zinc ores which is toxic to almost all systems and small doses can be dangerous. It is used in batteries, electroplating and PVC.

calcium cycle

The circulation of calcium atoms brought about mainly by living things. Thus calcium is taken up from the soil by trees and other plants and deposited in roots, trunks, stems and leaves. Rain may leach some calcium from the leaves and return it to the soil. Creatures such as insects, rabbits and other herbivores obtain their share of calcium from the plants and leaves; birds acquire it by eating the insects. Animals and birds die, leaves and branches fall and decay, and thus the calcium component returns to the soil. Some calcium may be lost from an ecosystem by leaching and surface run off carried to bodies of water. It is recycled through phytoplankton, zooplankton, fish and lake and ocean water.

Cambridge Water Company case

A celebrated case relating to liability for contamination of groundwater. In *Cambridge Water Co v Eastern Counties Leather plc* [1994] 1 All ER 53 the House of Lords dismissed an action brought by the statutory water company whose borehole for abstracting drinking water for public supply became unusable for that purpose following the discovery in it of organochlorines (trichlorethene and perchloroethene) at levels exceeding those permitted under EC Directive 80/778 concerning drinking water. On investigation, the source of the contaminants was shown to be the defendants' tannery, where there had been, prior to 1976, occasional spillage of solvents used in the tanning process. This had led to a gradual and cumulative leaching of the chemicals into the aquifer *(qv)* supplying the plaintiffs' borehole. The House of Lords reversed the decision of the Court of Appeal, and held that although in principle the defendants were subject to strict liability *(qv)* for the escape of the chemicals from the land (in the sense that liability would arise even if the defendants had exercised all due care to prevent such an escape), strict liability only arose if the defendants knew or ought reasonably to have foreseen that the chemicals might, if they escaped, cause damage. Since they could not reasonably have foreseen at the relevant time that the seepage of the solvent through their tannery floor could have polluted the water drawn from the defendants' borehole, they were not liable.

capillary action

Capillarity. A general term of phenomena observed in liquids due to unbalanced inter-molecular attraction at the liquid boundary, eg the rise or depression of liquids in narrow tubes, the formation of films, drops, bubbles etc.

carbon cycle

The biological cycle of carbon which results in the amount of carbon dioxide in the atmosphere remaining relatively constant. By a process of photosynthesis, plants take in carbon dioxide, absorb carbon and return oxygen to the environment. Oxygen is taken in by animal life and is combined with carbon from ingested plants to form carbon dioxide which is exhaled. Carbon dioxide thus released is again broken down by photosynthesis into carbon and oxygen and the cycle begins again.

carbon dioxide

A colourless, odourless, unreactive gas formed by, *inter alia,* the combustion of carbon and the respiration of animals. An important greenhouse gas *(qv).*

carbon monoxide

A poisonous, colourless, odourless gas, resulting from the incomplete combustion of carbon.

carbon tax

A tax levied on the use of fossil fuels and intended to reduce emissions of carbon dioxide from their inefficient combustion. The European Commission has proposed a Directive introducing a tax on carbon dioxide emissions and the use of energy (COM (92) 226) as part of a strategy to reduce greenhouse gas emissions and encourage the use of less polluting, more efficient or renewable energy sources. Under the proposed Directive the use of a number of specified fuels would be taxed on the basis of their respective carbon dioxide emissions as well as their energy (carbon) content. See also *economic instrument.*

carcinogen

A substance which causes cancer.

Carriage of Dangerous Goods by Road and Rail (Classification, Packaging and Labelling) Regulations 1994

Regulations (SI 1994 No 669) governing the carriage of certain potentially dangerous goods. As to the supply as opposed to the conveyance of such goods, see also *Chemicals (Hazard Information and Packaging for Supply) Regulations 1994.*

catalytic converter

A device fitted in the exhaust systems of motor vehicles running on unleaded fuel to reduce emissions of pollutants, particularly unburnt hydrocarbons, carbon monoxide and nitrogen oxides, by catalyzing chemical reactions to trap the pollutants. The fitting of three-way catalysts in new cars is required by EC Directive 91/441.

catchment area

The area draining naturally to a given point.

cause

The offence based on causing or knowingly permitting, for example under s.85 of the Water Resources Act 1991 (WRA 1991) 'causing or knowingly permitting any poisonous, noxious or polluting matter or any solid waste matter to enter any controlled waters', has from time to time been the subject of interpretation by the courts. 'Causing' involves an active operation which results in pollution and 'knowingly permitting' involves a failure to prevent pollution accompanied by knowledge. The word 'cause' does not imply an intention to pollute waters or that the prosecution must show negligence. The offence of causing water pollution or waste materials to be deposited on an unlicensed site is an offence of strict liability. It has been established, however, that an offence will not be committed when the behaviour of the accused is passive. Nevertheless the courts appear to be taking a more strict line in the interest of the protection of the environment.

In *CPC (UK) Ltd v National Rivers Authority* [1995], Environmental Law Reports, 1972 (D8) the polluting liquid had leaked from piping in CPC's factory, which had been defectively installed by sub-contractors before CPC had bought the factory, into the controlled waters. The defect was, therefore, latent. CPC argued that the cause of the pollution was not anything which it had done, but was rather the fault of the sub-contractors. The Court of Appeal rejected that argument saying that it was based on a false premise, namely that there could only be one cause of such an incident. The question to address was whether CPC could be said to have caused the escape and the answer was affirmative. The mere fact that the defect was latent was not relevant as s.85 WRA 1991 did not require fault or knowledge on the part of CPC nor did the section require proof that the defendant's act was the sole cause of the polluting incident. Consequently the fact that it could have been caused by some other person was irrelevant.

certificate of lawful existing use or development (CLEUD)
A certificate to the effect that specified development *(qv)* that has already occurred (eg the construction of buildings or works, or the change in use to the existing use) on specified land is lawful. It is granted under the Town and Country Planning Act 1990, s.191, and confirms the lawfulness of that development. There is a parallel procedure for certification where proposed development does not require planning permission: see also *certificate of lawful proposed use or development (CLOPUD).*

certificate of lawful proposed use or development (CLOPUD)
A certificate that specified proposed development may lawfully be carried out without further planning permission, having regard to whether it constitutes development at all, to any existing planning permission, lawful use or development; and to the Use Classes Order *(qv)* and the General Development Order *(qv)*. Granted under the Town and Country Planning Act 1990, s.192.

certificate of technical competence (COTC)

A certificate issued by the Waste Management Industry Training Advisory Board (WAMITAB) *(qv)* to persons satisfying the requirement specified under the Environmental Protection Act 1990, s.74(3)(b). The requirement is that an applicant for a waste disposal licence must show that the management of authorised waste disposal activities will be in the hands of a technically competent person. If an applicant is unable to show this he will not be a fit and proper person *(qv)* for the purposes of the Act and the application must be refused. The certificate is awarded on the basis of field assessment rather than written examination.

Chartered Institute of Water and Environmental Management (CIWEM)

The leading professional and qualifying body in the field of water and environmental management. CIWEM is a multi-disciplinary professional and examining body for engineers, scientists and other professionals engaged in water and environmental management. Membership is by examination and the present membership is over 11,500, of whom more than 9,000 are resident in the UK and the Republic of Ireland. The main objectives of CIWEM are to advance the science and practice of water and environmental management for the public benefit and to promote education, training, study and research in these areas. Its interests cover planning, design, construction, operation, maintenance, research and education in water, including water resources; flood alleviation; water supply and distribution; sewerage and sewage treatment; management and engineering of solid wastes and land reclamation and environmental conservation. CIWEM organises conferences and seminars, and publishes a journal newsletter and technical manuals on all aspects of water and waste water treatment.

chemical

A substance obtained by or used in a chemical process.

Chemical Industries Association (CIA)

An association established to represent the interests of the chemical industry and promote co-operation within the industry. The Association has devised a Responsible Care Programme which commits the industry to continuous improvement in all aspects of health, safety and environmental protection, and adherence to which is a condition of membership of the Association. [*Address: Kings Buildings, Smith Square, London, SW1P 3JJ; T: 0171 834 3399; F: 0171 834 4469.*]

chemical oxygen demand (COD)

The amount of oxygen used in the chemical oxidation of the matter present in a sample by a specific oxidising agent under standard conditions. See also *biological oxygen demand (BOD)*.

Chemical Release Inventory (CRI)

An inventory of industrial pollution emissions arising from industrial plants regulated under integrated pollution control (IPC) *(qv)* in England and Wales. The inventory is published by Her Majesty's Inspectorate of Pollution (HMIP) and lists substances emitted broken down by substance type, industrial sector and local authority area.

Chemicals (Hazard Information and Packaging for Supply) Regulations 1994 (CHIP)

Regulations that require suppliers and consigners of chemicals to classify the chemicals they supply by identifying their hazards and dangers. Where dangerous chemicals are supplied, suppliers are required to give information about the hazards to the person they supply and to package the chemicals safely. The 1994 Regulations (SI 1994 No 3247) came into force on January 31, 1995, and replaced the 1993 Regulations. They implement a large number of directives which amend and adapt technical progress to EC Directive 92/32 on the Classification, Packaging and Labelling of Dangerous Substances, and EC Directive 88/379 on the Classification, Packaging and Labelling of Dangerous Preparations.

The basic system of classification, packaging and labelling of chemicals for supply remains broadly the same, but:

1 provisions now relate to the European Economic Area;

2 advertisements for substances which are dangerous in supply must refer to the hazards presented by the substance; and

3 child-resistant fastenings must now be provided on the packaging of certain substances and preparations. The regulations deal only with supply. The carriage of dangerous substances is now covered by the Carriage of Dangerous Goods by Road and Rail (Classification, Packaging and Labelling) Regulations 1994 *(qv)*.

Chief Inspector's Guidance Notes (CIGN)

Guidance issued by the Chief Inspector of Her Majesty's Inspectorate of Pollution (HMIP) *(qv)* intended for use by inspectors in the assessment of applications for authorisation under the Environmental Protection Act 1990, s.6 (integrated pollution control). Each Note refers to a defined set of processes prescribed for integrated pollution control (IPC), and contains information about the processes, about relevant prescribed substances and their potential sources of release, together with guidance on best available techniques not entailing excessive cost (BATNEEC) *(qv)*. Release levels specified in the Notes are based on what is generally considered achievable by applying techniques which are appropriate to the sector of industry described. Inspectors then remain responsible for determining the actual release limits incorporated into the authorisations which must take account of BATNEEC and best environmental option (BEO) *(qv)*.

chlor-alkali process

A long-established process for manufacturing inorganic chemicals, in which brine (ie sodium chloride solution) is decomposed by electrolysis to produce chlorine gas, hydrogen gas and sodium hydroxide. Cheshire is the traditional centre of the UK chlor-alkali industry.

chlorination

The use of chlorine for the treatment of water. In waterworks, chlorination is used to kill bacteria and for the removal of algae growth, iron, manganese and sulphides. In swimming pools also it is used to control bacterial contamination, inhibit algae growths and improve water quality. In power generation, the chlorination of condenser cooling water helps to control slime-forming bacteria and to eliminate fouling by mussels, thus maintaining condenser efficiency. In sewage treatment works, chlorination may be used to improve final effluent quality.

chlorine

Halogen found naturally in combined form, usually as a chloride of potassium or sodium. Chlorine has poisonous properties and is often used in bleaching and purification of water. Chlorine has been identified as a potential ozone-depleting substance.

chlorine loading

The total amount of chlorine *(qv)* present in the atmosphere which gives an indication of the potential for damage to the ozone layer.

chlorofluorocarbons (CFCs)

A group of stable, man-made gaseous compounds used widely in refrigeration and air-conditioning systems, aerosol foam sprays, fire protection equipment and building insulation. CFCs have been identified as ozone-depleting substances and under the 1987 Montreal Protocol *(qv)*, 132 signatory nations committed themselves to reducing the production and use of CFCs (as well as other ozone-depleting substances). EC Regulation 594/91 on substances that deplete the ozone layer (as revised) sets a timetable for phasing out CFC production in Member States, and prohibits exports from the EC to non-member states.

chromatography

The separation of individual dissolved components in a solution (or a gas) by passing the solution or gas through a material (often held in a tube or column) that allows some components to travel more rapidly than others, depending on their molecular size, shape and polarity. Each component is eluted separately from the chromatographic medium and can be detected and measured. There are different methods such as gas-liquid chromatography (GLC or GC), high performance liquid chromatography (HPLC) and thin layer chromatography (TLC), all with different applications for chemical analysis.

chromium

A hard, white metallic element that occurs as chrome-iron ore, and is used in the manufacture of stainless steel.

Citizens' Environmental Initiative

An initiative launched by the United Kingdom Government in 1994 to encourage the growth of interest in the issues of sustainable development, so that personal choices could play a part in delivering sustainable development and to enlist the public's support and commitment. Since renamed 'Going for Green'.

civil liability

Liability arising under civil law as a consequence of an act or omission. For liability to attach, the plaintiff must establish, on the balance of probability, loss suffered of a legally recognised kind as a consequence of a breach of an obligation owed to him by the defendant. The loss or damage is typically redressed by the award of damages to compensate for the loss suffered, or by an order of the court such as an injunction (granted to restrain a person from committing a particular act) or specific performance (granted to direct that a particular act should be undertaken).

Civil liability provides a means for recovering damages for redressing environmental harm to persons or property, and in the environmental context the appropriate action is normally one or more of the common law torts of negligence, nuisance *(qv)*, trespass and *Rylands v Fletcher (qv)*. Statutes also create rights of civil action, eg the Environmental Protection Act 1990, s.73(6), which allows civil action to be brought against anyone who commits the criminal offence under s.33 of, *inter alia*, permitting the unauthorised disposal of waste, where the plaintiff has suffered damage as a result. Civil liability usually arises from the enforcement of private rights (which may indirectly benefit the environment, but are often therefore limited to the owned environment), but legislation sometimes also confers comparable rights on environmental agencies to sue polluters for costs incurred by the agency in preventing or cleaning up pollution (eg Water Resources Act 1991, s.161). See also *environmental liability*.

Clean Air Act 1993

The principal legislation controlling the emission of dark smoke from premises and chimneys, which was consolidated into one Act out of the Clean Air Acts 1956 and 1968. The 1993 Act confers powers on local authorities to designate smoke control areas within which it is prohibited to emit smoke from a chimney or to acquire or sell unauthorised fuel. Powers are given to the Secretary of State for the Environment to regulate specific forms of air pollution, for example from vehicle fuels and oil used for furnaces and engines.

clean-up costs

Expenditure incurred in remedying environmental damage. In some instances, statute confers a right on a public body to recover, from any person who caused or permitted pollution, expenditure reasonably incurred by that body in cleaning up or preventing environmental damage from it. Examples include: Environmental Protection Act 1990, s.59 (waste); s.81 (nuisance); and the Water Resources Act 1991, s.161 (pollution of controlled waters).

climate change

See *greenhouse effect.*

Climate Change Convention

The framework convention that was adopted at the Rio Summit *(qv)* in 1992, aimed at controlling emissions of greenhouse gases *(qv)*. The United Kingdom signed the Convention at Rio, and ratified it in December 1993; and it was also formally approved by the European Communities (EC) in December 1993 (94/69), but the EC has not yet supplied the necessary funds.

The objective of the Convention is the stabilisation of greenhouse gas concentrations in the atmosphere at a level that would prevent dangerous anthropogenic interference with the climatic system, within a time frame sufficient to allow ecosystems to adapt naturally to climate change, in order to ensure that food production is not threatened and to enable economic development to proceed in a sustainable manner. The Convention obliges developed countries who become parties to it to take measures aimed at reducing emissions of greenhouse gases to 1990 levels by the year 2000. The United Kingdom's programme for compliance is set out in *Climate Change: The UK Programme* (Cm 2427; 1994), which focuses on a partnership approach to energy conservation and the use of methane for energy generation.

clinical waste

Any waste *(qv)* which consists wholly or partly of human or animal tissue, blood or other body fluids, excretions, drugs or other pharmaceutical products, swabs or dressings, or syringes, needles or other sharp instruments, which unless rendered safe may be hazardous to a person who comes into contact with it; or any other waste arising from medical, nursing, dental, veterinary, pharmaceutical or similar practice, investigation, treatment, care, teaching or research, or the collection of blood for transfusion, which may cause infection to a person who comes into contact with it (Collection and Disposal of Waste Regulations 1988 (SI 1988 No 819) (as amended by the Controlled Waste Regulations 1992)).

Technical guidance on the meaning, classification and management of clinical waste is contained, *inter alia*, in Waste Management Paper No 25. The London Waste Regulatory Authority (LWRA) *(qv)* has led in the definition of codes of practice for the proper management and engineering of clinical waste. Further reference: *Guidelines for the Segregation, Handling, Transport and Disposal of Clinical Waste*, 2nd Edition, LWRA, Hampton House, 20 Albert Embankment, London SE1 7TJ.

Club of Rome

The authors of an important book, *The Limits to Growth* (1972), which examined five basic factors that determine and therefore ultimately limit growth on this planet, namely population, agricultural production, natural resources, industrial production and pollution. They predicted widespread disaster if fundamental environmental issues were not addressed: concerted international measures and joint long-term planning would be necessary and on a scale and scope without precedent.

coal

The fossilised carboniferous remains of plants commonly used as fuel. Coal is mined using both deep and opencast mining methods. There are a number of different types of coal, classed on their hardness and calorific value, eg anthracite, lignite. As with other fossil fuels the combustion of coal releases carbon dioxide, which is a greenhouse gas *(qv)*. Many coals also contain quantities of sulphur compounds which may also be released as gaseous oxides upon burning. These may cause the acidification of rainwater.

Coalition for Environmentally Responsible Economies (CERES)
An organisation in the United States formed by the Social Investment Forum (an organisation of investment and research professionals involved in socially responsible investments) and leading environmental groups, trade unions, churches and private and public bodies with investment duties. The guiding principles of CERES are the Valdez Principles *(qv)* which set broad standards for evaluating the effects of corporate activities on the environment.

coastal protection
The taking of measures against the erosion of land and other encroachments of the sea, such as sea walls, rock revetments and groyne systems. A responsibility of maritime district councils under the Coast Protection Act 1949. Securing defence from flooding of land by the sea is a responsibility of the National Rivers Authority under the Land Drainage Act 1991, except where sea defences are privately or local authority owned.

cocktail effect
The synergistic effect of more than one pollutant in the environment where the aggregate effect is greater than the effect of each pollutant taken separately.

code of practice
A set of rules without direct legal effect, commonly intended to promote best practice, and adopted by a Minister or firm or industry. In certain environmental contexts, such codes do not in themselves give rise to criminal liability *(qv)*, but they are admissible in evidence in assessing the accused's compliance with a statutory duty: examples include the *Code of Practice on the Waste Management Duty of Care* adopted under the Environmental Protection Act 1990, s.34; and an approved code of practice adopted under the Health and Safety at Work Etc. Act 1974, s.16.

co-disposal
The practice of the joint, controlled disposal of both biodegradable/putrescible *(qv)* municipal or commercial wastes (ie non-hazardous) with difficult industrial wastes (ie hazardous or toxic) in a landfill *(qv)*. Although not officially widely practised elsewhere, the practice of co-disposal has been developed on a scientifically

engineered basis by the UK over many years, and its continuance is authorised under a draft EC Directive on landfill although restrictions look likely to limit it to around half the 300 sites where it has been practised in the past. Further reference: HC Environment Committee, *The EC Draft Directive on the Landfill of Waste* (7th Report, Session 1990–91; HC 263); waste management paper *(qv)* 26F *Landfill Co-disposal* (issued by the DOE for public consultation in October 1994).

coliform

A type of bacteria found in sewage, and also in decaying vegetation. Limits to the presence of total coliforms and faecal coliforms are imposed by the following EC Directives: 75/440 on the quality of surface water required for abstraction of drinking water in Member States (Annex II); 76/160 on the quality of bathing water (Annex); 79/923 on the quality required of shellfish waters (Annex); 80/778 on quality of water intended for human consumption (drinking water directive), (Annexes I, II and III). Coliform analysis is undertaken by the National Rivers Authority in accordance with its manual, *Manual of Standard Methods for Microbiological Analysis* (1992). Proposals by the European Commission in 1994 for revision of the Bathing Waters Directive include a reduction in the total coliforms parameter, which is no longer thought to be a particularly useful indicator of sewage pollution, but retain the standards for faecal coliforms (*E. coli*).

collateral warranty

An undertaking, usually unilateral, entered into by one party in favour of another and collateral to a main contract, incorporated in a document which is often a deed. An example would be a warranty by an environmental consultant retained by client X to conduct a site investigation in favour of a purchaser, funder or some person, not a party to the main contract but relying on the consultant's report, to the effect that all appropriate care and skill has been exercised.

combined cycle gas turbine (CCGT)

An electricity generating station which uses natural gas as a direct and an indirect source of energy. As a direct source of energy gas is used to drive a gas turbine. The heat generated from this process is then used to drive a steam turbine. See also *combined heat and power*.

combined heat and power (CHP)

CHP power stations are stations which generate energy from the combustion of energy-rich materials but which also reuse some of the waste heat in secondary schemes, eg heating water for use in municipal heating schemes.

command and control

An approach to Government regulation of private activity which underlies most contemporary environmental law, in which the State sets rules and standards which are backed by criminal sanctions. As a philosophy, command and control is characterised by a faith in the achievement of public policy ends through coerced compliance. Legislation prohibits the undertaking of specified activities except in accordance with a licence granted by a regulatory agency, and the agency has enforcement powers, including a power to bring criminal proceedings. Examples in Britain include integrated pollution control *(qv)*, the licensing controls over discharges into the water environment (Water Resources Act 1991, s.85) and the waste management licensing regime under the Environmental Protection Act 1990, Part II. Critics of command and control point to the tendency for such agencies to be 'captured' by the very interests they are supposed to be regulating; and also to the potential economic inefficiency of imposing uniform standards across firms with differing abatement costs. See also *economic instrument; tradable permits*.

Commission for Sustainable Development

A United Nations Commission set up in 1993 following the United Nations Conference on Environment and Development (UNCED) *(qv)* to monitor progress in implementing the agreements made at the conference.

common agricultural policy (CAP)

The agricultural policy of the European Communities (EC), which provides significant financial and other support to agriculture, and which therefore has important environmental impacts. The 'greening' of the CAP was one of the objectives of the historic reform package agreed in 1992, which included the Agri-environmental Regulation *(qv)* and a Regulation (2080/92) on forestry measures. None the less, there is concern that the reforms are poorly funded, and that they do not address the problem of continuing funding under mainstream CAP programmes of farming practices which damage wildlife and environmental interests.

competent authority

The Government or other agency so designated by a Member State for the purposes of EC legislation. For example, EC Directive 85/337 on environmental assessment requires the designation of competent authorities for the purposes of authorising projects which are subject to environmental assessment *(qv)* (in Britain, local planning authorities and the appropriate Secretaries of State); and for water directives the competent authority for England and Wales is the National Rivers Authority.

compliance cost assessment (CCA)

An assessment of the costs likely to be imposed by new legislation, which is now required by the United Kingdom Government to be undertaken for new primary legislation (where a summary is given in the Memorandum to a Bill when it is introduced) and for statutory instruments (where it is placed on deposit for public inspection at the offices of the sponsoring Government department).

composting

The degradation of organic wastes in the presence of oxygen (ie aerobically). Composting is the first stage of degradation in landfill sites, where it utilises air trapped in the waste when landfilled; but it normally ceases fairly quickly as air is deliberately excluded. Organic wastes, particularly 'green' wastes, are often composted as a means of disposal, being converted into a soil conditioner which can be used to replace peat. Composting is sometimes used for the pretreatment of waste prior to landfill, to reduce the pollutant loading in leachates produced. The Draft Waste Strategy for England and Wales (January 1995) takes as an action point that 75% of local authorities will actively promote composting by the year 2000.

Comprehensive Environmental Response, Compensation and Liability Act 1980 (CERCLA)

The United States federal statute which established Superfund *(qv)*, under which special taxation revenues are paid to a fund for the clean-up of contaminated sites, and which empowers the Environmental Protection Agency to clean up the worst sites and recover contributions from parties who contributed to the contamination (42 U.S.C. 9601 et seq as amended by the Superfund Amendments and Reauthorisation Act 1986 (SARA) *(qv)*).

comprehensive general liability (CGL)

An American type of insurance policy which is the equivalent to public liability policies *(qv)* in the United Kingdom. This kind of policy covers the insured for legal liability incurred in the course of business in relation to injury or damage caused to people or to property. Between 1970 and 1984 the standard wording of these policies included a pollution exclusion clause, except for pollution arising from 'sudden and accidental' events.

conservation

The process of protecting and preserving, used in the environmental context in relation both to the built environment (see *conservation area*) and the natural environment, as in nature or species conservation, and habitat conservation.

conservation area

An area of special architectural or historic interest, designated under the Planning (Listed Buildings and Conservation Areas) Act 1990, for the purpose of preserving or enhancing its character or appearance. The power was introduced in 1967, and there are now over 7,000 conservation areas in England and Wales. Within a conservation area it is unlawful to demolish any building (some exceptions are prescribed) except in accordance with conservation area consent granted under the 1990 Act; or to fell or lop a tree unless prior notification has been given to the local planning authority and they have not made a tree preservation order. Policy guidance for conservation areas is in PPG15, *Planning and the Historic Environment* (1994). A distinctly different designation from a *special area of conservation* (SAC) *(qv)* which is a nature conservation designation under the EC Habitats Directive *(qv)* and the Conservation (Natural Habitats, &c.) Regulations 1994.

consignment note

In the environmental context this term describes the documentation used when waste materials are transported. It is applied to the system of disposal and transport for special waste *(qv)* under the Control of Pollution (Special Waste) Regulations 1988; and the transport of any waste under the EC Regulation on the supervision and control of shipments of waste within, into and out of the European Communities (259/93).

contaminated land

Land which because of its previous or current use has substances under, on or in it which, depending upon their concentration and/or quantity, may represent a direct potential or indirect hazard to man or to the environment. A badly neglected area of environmental damage until the late 1980s, when it came under the spotlight of the House of Commons Environment Committee in its reports on *Toxic Waste* and *Contaminated Land*. There is no standard definition of contaminated land, and estimates therefore vary widely as to the number of sites which should be regarded as contaminated, and which requires some measure of clean-up. Evidence to the House of Commons Select Committee (*Contaminated Land*, para. 24) suggested that there might be as many as 100,000 contaminated sites in the United Kingdom. Contaminated land may require remedial work to prevent the occurrence of environmental damage, such as the removal or treatment of soil, the monitoring and flaring of methane *(qv)*, or the physical containment of the contaminating substances through bunding and other means.

Treatment of contaminated land is expensive, and in some cases the clean-up cost exceeds the market value of the land after clean-up. This remains a controversial area of public policy, which has in the United States led to the Superfund *(qv)* scheme, and in Europe there is currently considerable interest in introducing new liability rules (see *environmental liability*). There are no reliable estimates of the number of contaminated sites in the United Kingdom that require clean-up. It is widely accepted that paying for clean-up requires contributions from general Government revenues, from those who caused the pollution and from the present owners of the land.

A new regime for contaminated land in the United Kingdom is proposed in the Environment Bill 1995, which defines it as land which, in the opinion of the district or unitary council, is by reason of substances in, on or under it, causing or is likely to cause, harm or pollution of controlled waters. 'Harm' means harm to the health of living organisms or other interference with the ecological systems of which they form part and, in the case of man, includes harm to his property. Local authorities are given a new function of identifying contaminated land, and are required to report to the new Environment Agency on any sites affecting controlled waters, and any closed landfill sites.

The local authority become the 'enforcing authority' in respect of these sites. They are required to seek the Agency's advice in relation to any site from which there is, or is likely to be, pollution to controlled waters. They must also notify the Agency of any closed landfill sites in their area. Where a closed landfill presents risks of serious harm, or serious pollution of controlled waters, the Secretary of State may designate it as a site for which the Agency will actually become the enforcing authority. The enforcing authority are required to prepare and publish a statement specifying what is to be done to remediate a closed landfill site, and may secure that remediation by serving a notice on the appropriate person setting out what reasonable action needs to be taken. The appropriate person is, first, the person who caused or knowingly permitted the contaminating substances in question to be there; and secondly, if such person(s) cannot be found, or has transferred liability to the occupier or owner of the land for the time being, then the owner or occupier for the time being of the land. The notice may specify only those works which the enforcing authority think reasonable, having regard to the cost likely to be involved and the seriousness of the harm or pollution in question.

The enforcing authority are to have regard to guidance issued by the Agency as to what a remediation notice may require a person to do, and the standard to which any land or waters are to be remediated. There is a right of appeal against a remediation notice, and it is an offence not to comply with a notice that has taken effect. The Agency is entitled itself to carry out the necessary works in default, and to charge the cost back to the appropriate person, though subject to consideration of whether hardship would result if full recovery were to be pursued. Further reference: NRA, *Contaminated Land and the Water Environment* (Water Quality Series No 15; 1994); S. Tromans and R. Turrall-Clarke, *Contaminated Land* (Sweet & Maxwell, 1994).

contaminative use

An expression used in the Environmental Protection Act 1990, s.143, which defines it as 'any use of land which may cause it to be contaminated with noxious substances'. That section required local authorities to establish and maintain registers of land which might be contaminated, and allowed the Secretary of State to specify contaminative uses of land for the purposes of identifying which land should go onto such a register. The provisions proved highly controversial, and the Department of the Environment (DOE) consulted twice on proposals for setting up the registers before eventually abandoning the scheme altogether. Section 143 was never brought into force, and its repeal is proposed by the Environment Bill 1995.

Contract Regarding a Supplement to Tanker Liability for Oil Pollution (CRISTAL)

An international voluntary regime through which compensation for clean-up costs and pollution damage is available after oil spills from oil tankers. Set up to complement international conventions relating to oil pollution damage.

controlled waste

Waste which is controlled under the Environmental Protection Act 1990, Part II, for which purposes it is defined (by s.75(4)) as 'household, industrial and commercial waste or any such waste' (meaning any combination of household, commercial or industrial waste). Each of those headings is separately defined by s.75 and by regulations made under that section, with the effect that controlled waste does not include mine and quarry waste, waste from agricultural premises, certain sewage sludge or radioactive waste. To a large extent the definition of controlled waste has been superseded by that of directive waste because of the advice given in DOE Circular 11/94.

controlled waters

Waters which are subject to the abstraction *(qv)*, impoundment *(qv)* and pollution control regimes under Part III of the Water Resources Act 1991, and which include territorial waters (up to three miles offshore), coastal waters, surface water (ponds, lakes and rivers, including those which are artificial) and groundwater *(qv)*.

Control of Major Accident Hazards Regulations 1984 (CIMAH)

Regulations which aim to prevent and limit the potential for and effects of accidents from specified industrial activities which involve dangerous substances (usually in large quantities). The Regulations implement the EC Directive 82/501 on Major Accident Hazards of Certain Industrial Activities. This Directive is also commonly referred to as the Seveso Directive.

Control of Pollution Act 1974 (COPA)

A major step in the development of British environmental legislation. The Bill was introduced to Parliament initially by the Conservative Government as the Protection of the Environment Bill in 1973, and revived largely unamended by the incoming Labour administration in February 1974. Although its major function was to consolidate various previous measures, it also introduced a new regime for the control of the deposit of waste on land, which became the model for the EC Framework Directive on waste in 1975. It adapted with significant amendments the former controls over river pollution, to coincide with the creation of the new water authorities under the Water Act 1973 (although implementation of many of the new provisions, notably those relating to public registers of discharge consents, was then delayed for several years). The Act also provided new administrative powers in respect of noise abatement (Part III); and laid down a new framework for the control of emissions from motor vehicles (Part IV). Its provisions are now almost entirely superseded by the Environmental Protection Act 1990 and the Water Resources Act 1991.

Control of Substances Hazardous to Health Regulations 1994 (COSHH)

Regulations (SI 1994 No 3246) which impose requirements in respect of the management of the occupational exposure of employees to substances which are likely to be hazardous to health. The latest Regulations superseded those of 1988. The new Regulations reflect the creation of the European Economic Area and also implement Directive 90/679 on the protection of workers from risks related to exposure to biological agents at work by applying the Regulations to such agents and supplementing their provisions with a Schedule containing special provisions relating to such agents. Biological agents are defined to mean any micro-organism, cell culture, or human endoparasite, including any which have been genetically modified, which may cause any infection, allergy, toxicity or otherwise create a hazard to human health. The basic duty is to prevent (or where prevention is not practicable, to control) the exposure of employees to hazardous substances. The list of substances which are defined as hazardous to health is wide ranging and includes substances which are toxic, corrosive or irritants, dust, micro-organisms, and substances for which a maximum exposure limit is specified in the Regulations. Employers are required to undertake a risk assessment of their own, and to make it available for inspection by the statutory authorities. The Health and Safety Executive (HSE) *(qv)* has undertaken a number of prosecutions for the absence or inadequacy of COSHH. See also *maximum exposure limit (MEL); occupational exposure standard (OES).*

Convention on Biological Diversity

One of the conventions agreed at the United Nations Conference on Environment and Development (UNCED) *(qv)* and signed by over 150 countries, including the UK and the EC. The Convention establishes a framework for action at the national level to use and conserve biodiversity *(qv)*. Under the Convention the UK and other contracting parties are required to develop national strategies, plans or programmes for the conservation and sustainable use of biological diversity, or to adapt existing strategies, plans or programmes for this purpose. *Biodiversity: the United Kingdom Action Plan* (Cm 2478; 1994) is part of the UK Government's follow up to the Convention. It commits the UK to sustainable use of biological resources and to use of non-renewable resources, and seeks ways of integrating the objectives of biodiversity conservation into Government policy and action, following the precautionary principle *(qv)*.

Convention on Environmental Impact Assessment in a Transboundary Context

A convention concluded at Espoo in Finland in 1991, under the aegis of the UN Economic Commission for Europe (ECE), which seeks to prevent and control transboundary environmental damage by establishing an agreed mechanism for assessment. The parties agree that, prior to authorising or undertaking a proposed activity listed in Appendix I (which corresponds to Annex I to EC Directive 85/337 on environmental assessment) that is likely to cause a significant adverse transboundary environmental impact, they will undertake an environmental impact assessment, and allow an opportunity for the public in the areas likely to be affected to participate in the process.

The Convention prescribes the procedure to be followed, and contains provision for arbitration between parties to it in the event of dispute.

Convention on International Trade in Endangered Species (CITES)
The Convention which was agreed in Washington on March 3, 1973 and which aims to protect endangered wildlife species by controlling international trade. The protected species are listed in the Convention, and international trade in those species which are threatened with extinction and are listed in Appendix I is prohibited. For certain other species which are not yet threatened with extinction but which may become so, and which are listed in Appendix II, international trade is permitted under a system of permits, but only so far as it does not threaten the survival of the species. The Convention was implemented in the European Communities by EC Regulation 3626/82, as subsequently amended.

Convention on the Transboundary Effects of Industrial Accidents
A convention (Cm 2443) signed in Helsinki in 1992 in the interests of protecting human beings and the environment against the effects of industrial accidents. It obliges the parties to take appropriate measures to prevent industrial accidents, to establish policies on the siting of new hazardous activities, to establish emergency preparedness, to involve the public in decision-making, to establish an industrial accident notification system, response measures and co-operate in mutual assistance. The Convention does not lay down liability rules. It has been signed, but not ratified, by the United Kingdom.

Convention on Wetlands of International Importance Especially as Waterfowl Habitat
See *Ramsar Convention.*

cooling tower
A device for cooling water by evaporation in the ambient air. A tower requires a flow of air and this may be induced by natural or mechanical means. The use of large cooling towers at power stations to dissipate recovered waste heat eliminates the problem of thermal pollution in masses of water. However, the water for a cooling tower system is taken from the river or stream, thus eliminating its subsequent use down-river. After being used several times for cooling purposes, it is finally dissipated to the general atmosphere as steam.

COREPER
The French acronym for the EC Committee of Permanent Representatives consisting of the 15 ambassadors from the Member States.

Coreper's position inside the Communities is unique. It has decision-making powers and its permanent presence in Brussels, alongside the Commission, gives it an extra edge. One of the best kept secrets in Brussels is that 90% of EC decisions, which include environmental decisions, are resolved informally in Coreper before they even reach national ministers.

cost-benefit analysis (CBA)

A formal evaluation of the respective costs and benefits of taking or not taking a particular course of action, which may attempt to attach monetary values to public and private benefits or disbenefits. A version of CBA is now formally required of the European Commission when proposing new environmental legislation (under Article 130r(3) of the Treaty), which requires it to take account of 'the potential benefits and costs of action or lack of action'; and a parallel duty is proposed for the Environment Agency and the Scottish Environment Protection Agency by the Environment Bill 1995, under which each will be required to take into account the costs which are likely to be incurred, and the benefits which are likely to accrue, before deciding whether and how to exercise any power. The duty does not arise where it is unreasonable in view of the nature or purpose of the power or the circumstances of the case; nor where the Agency is under another duty to comply with any requirements, or pursue or meet any aims or objectives.

Council for the Protection of Rural England (CPRE)

A non-governmental organisation concerned with the advancement and conservation of the rural environment in England. In particular the CPRE seeks to enhance the beauty and variety of the countryside by influencing decision-makers at all levels including the European Communities, Government and local authorities. [*Address: Warwick House, 25 Buckingham Palace Road, London, SW1W OPP; T: 0171 976 6433; F: 0171 976 6373.*]

Council of Europe

The first European political institution (founded in 1949), which today has 32 member countries. Its work was partially superseded and largely overshadowed by the establishment of the European Communities (EC), but it continues to have a role in safeguarding and realising the ideals and principles which are the common heritage of its member countries and in facilitating economic and social progress. The Convention on Civil Liability for Damage Resulting from Activities Dangerous to the Environment (the Lugano Convention) was concluded in 1993 under the Council's auspices, (not ratified by the United Kingdom).

Council of the European Communities

Previously known as the Council of Ministers, the Council is the most powerful organ of the European Communities (EC). Its primary task is to consider proposals for EC legislation and vote on whether proposals should become law. Each Member State has a representative on the Council who is accountable to its own executive. The identity of the representative will vary according to the subject under discussion. Thus for environmental matters the Council consists of the environment ministers from each Member State. Decisions are made by unanimous agreement, qualified majority voting or by simple majority depending on the subject matter. Presidency of the Council is held for a six-month term by each Member State in rotation.

Countryside Commission for England (CC)

An independent statutory agency created by the Countryside Act 1968 out of the former National Parks Commission that had been established under the National Parks and Access to the Countryside Act 1949. The Commission has responsibility for the conservation and enhancement of the beauty of the English countryside including making it more accessible for the purposes of public enjoyment. The Commission is responsible for designating national parks *(qv)* and areas of outstanding natural beauty (AONB) *(qv)*; defining Heritage Coasts; and establishing National Trails for walkers and riders. [*Address: John Dower House, Crescent Place, Cheltenham, Gloucestershire, GL50 3RA; T: 01242 521381; F: 01242 584270.*]

Countryside Council for Wales (CCW)

The Welsh counterpart for both the Countryside Commission for England (CC) *(qv)* and English Nature *(qv)*. [*Address: Plas Penrhos, Ffordd, Penrhos, Bangor, Gwnedd, LL57 2LQ; T: 01248 370444; F: 01248 355782.*]

county matter

An area of development control *(qv)* under planning legislation for which the powers are allocated to English and Welsh shire county councils rather than the districts; includes all development relating to minerals and (except in Wales) waste. The classification will be superseded in Wales from April 1, 1996 upon the establishment of unitary local government under the Local Government (Wales) Act 1994.

county site

Not a technical term, but commonly used to denote a site whose importance for habitat *(qv)*, protection is of less than national significance, and hence is not designated a site of special scientific interest *(qv)*; but is of more than purely local significance. Such sites are often listed in documents, not necessarily widely published, produced by County Wildlife Trusts, and may be noted in development plans *(qv)*.

cradle to cradle

A term becoming increasingly popular amongst industrial (eg automobile) designers which emphasises the necessity to produce a product or part of it, so that when its life is finished it can be returned to its original manufacturing source and recycled for similar use again.

cradle to grave

A term employed in the life cycle analysis (LCA) *(qv)* of a product, in which 'cradle' refers to the environmental impact of the extraction of raw materials used in a product and their distribution, and 'grave' refers to the environmental impact of the eventual disposal of the product. Also used in relation to the tracking of waste materials from their production source to their final interment or destruction. See also *producer responsibility*.

criminal liability

Liability to sanctions under criminal law (such as probation, fine or imprisonment) for the commission of a criminal offence. Criminal sanctions are commonly used to reinforce environmental controls (eg causing or knowingly permitting the unauthorised entry of poisonous noxious or polluting matter into controlled waters, under the Water Resources Act 1991, s.85). Breach of such a prohibition renders the offender liable to prosecution, normally by the regulatory agency concerned but private prosecution is also now possible for most environmental offences.

The prosecution must prove the offence beyond all reasonable doubt. In most cases, this involves proof both of a physical act or omission (the *actus reus (qv)*), and a mental state (the *mens rea*), but some environmental offences are regarded as imposing strict liability *(qv)* for which the prosecution need prove only the *actus reus*, on the theory that it is essential in the interests of the general public good to impose the highest standards of preventive behaviour on potential polluters.

critical load

The maximum pollutant loading that a given ecosystem can tolerate without suffering adverse effects.

critical natural capital

An aspect of sustainable development *(qv)*. Certain aspects of the environment which perform vital functions (eg greenhouse gases *(qv)*) and which are not capable, if lost, of being replaced by man-made capital.

cross-media

A term generally used to describe the propensity of pollutants to pass from one environmental medium to another or to affect more than one medium simultaneously. For example the scrubbing of acid gases from a combustion process may lead to the transfer of those materials in water which would then cause problems if discharged to the aqueous environment, hence transferring the problem from one environmental medium to another. The neutralisation of these scrubber liquids may lead in turn to precipitation and the production of a solid waste for disposal to land. It is the cross-media or trans-sectoral nature of many forms of pollution which has led to new institutional arrangements and control methods in the last decade, including integrated pollution control *(qv)*, Her Majesty's Inspectorate of Pollution *(qv)* and the proposed Environment Agency *(qv)*. Further reference: Royal Commission on Environmental Pollution *(qv) 3rd Report* and *5th Report.*

cultural heritage

A broad expression which includes all of today's cultural assets that have been inherited from the past, including music, sculpture, fine art, crafts and cultural traditions. In the environmental context its main importance is in relation to the protection of historic buildings and archaeological remains. There is an obligation to assess the impact of proposed development upon the cultural heritage when undertaking an environmental assessment *(qv)* under EC Directive 85/337 and under the 1988 British Regulations transposing its requirements.

The expression now also appears in the Treaty of Rome, to which the Maastricht Treaty added a new Title XI, 'Culture', which requires the Community 'to contribute to the flowering of the cultures of the Member States, while respecting their national and regional diversity and at the same time bringing the common cultural heritage to the fore'. The Community is required, *inter alia*, to 'take cultural aspects into account in its action under other provisions of this Treaty' (Article 128(4)), and the Council is empowered to 'adopt incentive measures, excluding any harmonisation of the laws and regulations of the Member States'. The Environment Bill 1995 *(qv)* specifies the protection and enhancement of the cultural heritage as an objective of the new National Park Authorities.

cyanides

Inorganic compounds which are the salts of hydrocyanic acid, and which exist in two chemical forms, as free or simple cyanides, and as complex cyanides. Certain types of ferrous engineering components have been, and to some extent are, given heat treatment by immersion in molten salt baths in order to confer on them certain properties such as case hardening. The use of cyanides began in the early 1920s and expanded in the early 1930s but has recently been significantly decreasing.

Hydrogen cyanide, its simple salts and their solutions present a high risk to man. A single dose of 100 mg of sodium cyanide can be fatal if taken orally whilst exposure to hydrogen cyanide gas in air is immediately fatal; it is rapidly absorbed through skin, eyes and mucus membranes. Under normal conditions, the body is capable of dealing with a small continuous burden of cyanide and a dose which would be lethal if taken in by the body as a single dose, may be tolerated without effect if absorbed slowly over a longer period. Further reference: *DOE Waste Management Paper No. 8,* (2nd ed) (currently under revision).

cyclone

1 A device for removing particulate matter from the waste gases of industrial processes. A simple cyclone consists of a cylindrical upper section and conical bottom section. The dust-laden gases enter the cylindrical section tangentially. Centrifugal action throws the grit and dust particles to the outer walls. These particles fall by gravity to the dust outlet at the bottom. The relatively clean gases leave through a centrally situated tube within the upper section. Cyclones may be used singly, or in groups or nests.

2 In meteorology a low pressure area with winds rotating counter-clockwise around the centre in the northern hemisphere, and clockwise in the southern hemisphere. Winds bring in moisture and rainy, windy weather prevails as rising air cools and vapour condenses.

damages

A sum of money awarded by a civil court to the plaintiff by way of compensation for wrong done. Exceptionally, but the normal function is purely compensatory, exemplary or punitive damages may also be awarded for purposes of punishment.

dangerous substance

In addition to its obvious colloquial meaning, has a specific meaning under the EC Directive (76/464/EEC) on pollution caused by certain dangerous substances discharged into the aquatic environment of the European Communities. The Directive lists certain substances which if discharged into water may be harmful or have a deleterious effect upon the aquatic environment. The most harmful substances are set out in List I of the Annex to the Directive (known as black list *(qv)* substances, eg mercury, cadmium, aldrin *(qv)*) and less harmful, but none the less harmful substances, appear in List II of the Annex (known as grey list *(qv)* substances).

Danish Bottles Case

The expression commonly used to identify the case of *Re Disposable Beer Cans: EC Commission v Denmark (Case 302/86)* [1989] 1 CMLR 619, where the European Court of Justice (ECJ) *(qv)* was called upon by the European Commission *(qv)* to determine the validity of a unilateral measure of environmental protection in circumstances where conflict arose with Treaty of Rome objectives relating to the removal of barriers to trade. Article 30 of the Treaty of Rome prohibits the introduction by Member States of measures which have as their effect restriction upon trade. Article 36 limits the applicability of Article 30 where it can be established that measures introduced are justified in terms of non-economic objectives which are also objectives recognised by the European Communities (EC), and provided that a measure so introduced is neither arbitrarily discriminatory, nor a disguise for primary and underlying objectives of an economic nature.

Between 1978 and 1981, Denmark had introduced controls over the packaging and marketing of beer and soft drinks, requiring them to be sold only in refillable bottles bearing a mandatory deposit. This was, in fact, the customary national practice for Danish lagers in Denmark. Metal cans were banned altogether, and the requirement that containers be reusable effectively prohibited use of plastic bottles. Containers had to be approved before use, in order to restrict the variety of bottles in circulation, and pave the way for the effective administration of a universal return and refund scheme.

The Court ruled that the restrictions on non-approved containers were in breach of Article 30 of the EC Treaty, but accepted the prohibition on metal containers, and the requirement of re-usability. In doing so, the Court held:

1 that environmental objectives constitute non-economic

objectives for the purposes of Article 36 which are at the same time 'mandatory' objectives of the European Communities. As measures purporting to have their basis in such objectives, the Danish measures were considered to meet this requirement of validity;

2 that measures introduced must not advantage national products or producers against those from other Member States. On the facts, a distinction was made here between the prohibition on metal containers and the requirement that containers be re-usable on the one hand, and the requirement for container approval on the other. The latter was considered to constitute a discriminatory measure;

3 measures introduced must be proportionate to the objective in view: where several methods are available to achieve a non-economic objective, the one least restrictive to the free movement of goods must be selected. On the facts, the ECJ again distinguished the prohibition and re-usability requirements from that relating to container approval. The former measures, notwithstanding their impact upon free trade, were considered in proportion to the non-economic objectives in view. The latter, however, imposed costs in terms of trade that could not be justified in terms of, and which were out proportion to, the non-economic benefits conferred.

daughter directive

A subordinate directive which fleshes out a specific area reserved by a main, or framework, directive. See also *directive*.

decibel

A unit for describing the ratio of two powers. It is the most commonly used unit for measuring sound levels by comparing them to a reference sound level. It is one tenth of a bel.

decibel (dBA) scale

An international weighted scale of sound levels which attenuates the upper and lower frequency content and accentuates middle frequencies, thus providing a good correlation in many cases with subjective impressions of loudness and sense of annoyance. Nearly all audible sounds lie between 0 and about 140 dBA. 0 dBA is the threshold of hearing, while sounds above 140 dBA are not common. An increase of 10 dBA means that the noise perceived by a listener has roughly doubled in loudness. A car passing at 70 dBA sounds twice as loud as one passing at 60 dBA.

Examples of typical dBA levels are:

dBA	
rustle of leaves	10
quiet office	50
busy office	65
moderate traffic	70
alarm clock	80
very noisy factory	90

deep pocket liability

See *joint and several liability; Superfund.*

de-inking

Series of processes by which various types of printing inks are removed from paper fibre pulp during the re-processing and re-cycling of recovered paper products. Particularly necessary where high quality and whiteness of the finished product are required.

deoxyribonucleic acid (DNA)

A complex molecule present in all living cells and which contains all the information for cellular structure, organisation and function.

Department of the Environment (DOE)

A Government Department for England (though with some strategic responsibilities in relation to the rest of the United Kingdom) responsible *inter alia* for environmental protection, countryside affairs, local government, housing, conservation and planning. It was first formed in 1972 from a merger between the Ministry of Public Buildings and Works, the Ministry of Housing and Local Government and the Ministry of Transport (which has since become a separate Department). Separate divisions within the Department have responsibility for environmental protection; local government and planning; and countryside and cities. For further details of current allocations of responsibility, and contact points, see the annual *Civil Service Year Book* (HMSO). [*Address: 2 Marsham Street, London, SW1P 3EB; T: 0171 276 3000; F: 0171 276 0818.*]

Department of Trade and Industry (DTI)

A Government Department for the United Kingdom, responsible for a wide range of international and national commercial issues. In the environmental context the Department plays a role in research and development in science and technology. In addition it has taken steps to provide and increase awareness of British business to opportunities relating to environmental protection and environmental technology as, for example, the Energy Technology Support Unit (ETSU) *(qv)*. For further details of current allocations of responsibility, and contact points, see the annual *Civil Service Year Book* (HMSO). [*Address: 123 Ashdown House, Victoria Street, London, SW1E 6RB; T: 0171 215 5000; F: 0171 215 5665 (library).*]

Department of Transport (DTp)
A Government Department responsible for land, sea and air transport. [*Address: 2 Marsham Street, London, SW1P 3EB; T: 0171 276 3000.*]

deregulation
The process of lifting regulatory and other controls over individual and, in particular, business activity. Has led to fears that environmental restrictions might be diluted, particularly under the Deregulation and Contracting Out Act 1994, s.5, which confers a broad power on any Minister of the Crown to make an order to improve (so far as fairness, transparency and consistency are concerned) enforcement procedures for any restriction, requirement or condition imposed by statute, so long as that would not jeopardise any necessary protection.

derelict land
Land which is so damaged by industrial or other development that it is incapable of beneficial use without treatment. Derelict land grant is payable under the Derelict Land Act 1982 to local authorities, other public bodies, private firms and individuals for the reclamation of derelict land, and can assist in works of clean-up of contaminated land *(qv)*.

desalination
The process of removing soluble salts from liquids, and used primarily to remove sodium chloride (NaCl) from water (H_2O) for the production of potable (drinking) water from sea water.

desertification
The process by which fertile land becomes desert, for example by erosion, drying up of aquifers or abnormal climatological conditions.

detergents
Surface active agents capable of removing dirt and grease from a variety of surfaces. The synthetic household detergents that were marketed some 40 years ago contained, as their principal component, an alkylbenzene sulphonate, which proved particularly resistant to decomposition by bacterial action, unlike soaps which undergo rapid decomposition. Thus, the early detergents were only

partially decomposed during the treatment of sewage and about half the quantity originally present was subsequently discharged with the sewage effluent, resulting in foaming. They impeded the rate through which oxygen could be transferred from the gas phase to solution, and so not only raised the cost of sewage treatment but also tended to inhibit the processes of self-purification in water courses. Later substitutes are more susceptible to bacterial attack.

development

For the purposes of development control *(qv)* under the Town and Country Planning Act 1990, defined as the carrying out of building, engineering, mining or other operations in, on, over or under land (so-called operational development); or the making of any material change in the use of any buildings or other land. It is generally unlawful to carry out development without planning permission, for which application must be made to the local planning authority *(qv)*, though the Town and Country Planning (General Permitted Development) Order 1995 (SI 1985 No 418) grants permission for an extensive range of activities (including afforestation and certain agricultural development) and the Town and Country Planning (Use Classes) Order 1987 allows certain changes of use to occur without planning permission.

development control

The function of controlling the development *(qv)* of land under town and country planning legislation. The statutory development plan for an area does not confer any entitlement to develop land. Planning permission is therefore required on each occasion that development is proposed and is not otherwise authorised. In determining an application for planning permission, the local planning authority *(qv)* are required to have regard to the development plan and to any other material considerations (Town and Country Planning Act 1990, s.70), and to determine the application in accordance with the plan unless material considerations indicate otherwise (s.54A). If permission is refused, or granted subject to conditions, the applicant is entitled to appeal to the Secretary of State for the Environment.

The system of control is discretionary, and the environmental implications of the proposed development are capable of constituting a 'material consideration' which can displace the policies in the development plan. Development control is the principal means through which the requirements of EC Directive 85/337 on environmental assessment is implemented in Britain, by treating planning permission as the authorisation required for potentially harmful projects under that Directive, and requiring that for projects within the Directive any application for planning permission should be accompanied by an environmental statement *(qv)*, upon which the authority are required to consult widely, and to take into account in determining the application. Similarly, development control provides the principal mechanism for protecting natural habitats from development, by virtue of the Conservation (Natural Habitats, &c.) Regulations 1994 (SI 1994 No 2176). Advice on the use of development plans and development control to attain environmental objectives is given in PPG9, *Nature Conservation* (1994), and PPG23, *Planning and Pollution Control* (1994).

development plan

The statement of local planning policies that each local planning authority *(qv)* is required by statute to maintain, and which can only be made or altered by following the procedures prescribed for that purpose, which include obligations to consult widely and to hold a public local inquiry into objections. The development plan includes:

1 the structure plan for the area (normally prepared by the county council); and

2 an area-wide development plan for each district council area. In London and the six metropolitan areas of West Midlands, Merseyside, Tyne and Wear, South Yorkshire, Greater Manchester and West Yorkshire, this two-tier structure is in the process of being superseded by new unitary plans prepared by the London boroughs and the metropolitan district councils. In areas where there is still no comprehensive local plan coverage, the old-style development plans prepared under pre-1968 legislation remain part of the formal development plan, though these are often now far removed from contemporary reality. In making or amending development plans, local planning authorities are required to have regard to environmental considerations, and PPG12, *Development Plans and Regional Planning Guidance* (1992), advises on the environmental appraisal that should be undertaken of the proposed development plan as a whole.

dialysis

The filtration of substances through semi-permeable membranes.

dichloro diphenyl trichloroethane (DDT)

A previously extensively used agricultural pesticide, which is stable under most environmental conditions and resistant to complete breakdown by soil micro organisms. Its use is now heavily restricted, and EC Directive 79/117 prohibits the sale of pesticides containing it. DDT is also a black list *(qv)* substance.

dieldrin

A white crystalline insecticide produced by oxidation of aldrin *(qv)*. It is a red list *(qv)* substance. It has been suggested that a drop in the breeding rate of Golden Eagles in Scotland was linked to the introduction of dieldrin into sheep dips, because of the occurrence of dieldrin in unhatched eggs. Dieldrin is now banned and the Golden Eagle breeding rate has returned to normal.

diffuse source pollution

Pollution which arises from various activities with no discrete source; non-point source pollution. See also *point source pollution.*

digestion

The biochemical decomposition of organic matter using anaerobic *(qv)* bacteria *(qv)* which results in the formation of simpler and less offensive organic compounds.

dilute and disperse

The practice of allowing leachates (rainwater contaminated by flow through wastes) to permeate into underground water-bearing strata, whereby it is diluted by large quantities of permeating water and allowed to spread out over a large area and thereby be rendered less concentrated or non-hazardous. Landfill sites may be divided into two broad categories according to the overall approach to the management of leachate *(qv)*: non-containment and containment sites. Non-containment sites do not attempt to prevent leachate from percolating to the environment and the prevention of pollution depends on the attenuation and diluting mechanisms operating both within the waste and in the strata, beneath and adjacent to the landfill site. Uncertainty about the attenuation mechanisms, however, precludes accurate prediction of the pollution risk. The phrase dilute and disperse, current in the late 1970s, changed gradually into dilute and attenuate. There is, in 1995, still only a developing understanding of the mechanisms by which leachate is attenuated and how methane is generated from waste. Research continues as to why and how all wastes degrade in a landfill site and during migration so that the safe boundaries of co-disposal, hitherto determined largely by empirical tests on leachate, can be more firmly established.

dioxins

Collective name for the group of chemicals called the dibenzo-*p*-

dioxins. Each of three unsaturated cyclic compounds, $C_4H_60_2$ and C_4H_40 and any derivative of such a compound, eg tetrachlorodibenzoparadioxin. Dioxins are produced during combustion of organic materials and wastes, for example during the manufacture and the incineration of certain plastics. They are very persistent in the environment due to their chemical stability. They are regarded as probable human carcinogens and may affect development, reproduction and the immune system. Current concern centres around emissions from the sintering plants of steelworks, from diesel combustion and from incineration, although the Royal Commission on Environmental Pollution *(qv)* in its 17th Report, *Incineration of Waste*, believed that new controls would lead to a steep fall in emissions.

direct effect

The doctrine that a directive *(qv)* of the European Communities (EC) may have legal effect in a Member State, despite the failure of the Member State properly to transpose it. The doctrine has been developed in decisions by the European Court of Justice, and proceeds on the basis that States should not be able to take advantage of their failures in transposition to avoid liability. The doctrine allows an individual to enforce terms of a directive against a Member State (or any emanation of it; see below) in its national courts if:

1 it is sufficiently unconditional and precise in its terms to grant rights to the individual;

2 the individual's rights have been infringed and he has suffered as a result of the failure to transpose the directive. EC Directive 85/337 was held in *Wychavon District Council v Secretary of State for the Environment* [1993] Env LR 239 not to satisfy this test. In order to constitute an 'emanation of the State' a body must be providing a public service; must be under the control of the State and must for that purpose possess special legal powers beyond those normally available to individuals under general law. A health authority is an emanation of the State for the purposes of the doctrine, and so too is a water and sewerage undertaker: *Marshall v Southampton Health Authority* [1986] 1 QB 401; *Griffin v South West Water Services Ltd* (High Court, August 25, 1994). See also *sewerage undertaker; water undertaker.*

directive

The principal type of European Communities (EC) legislation which has been used in environmental matters. A directive is binding upon each Member State as to the objectives to be achieved, but allows

Member States to adopt whatever form or method they deem appropriate to transpose those objectives into national law by a date specified in the directive. Some directives are framework directives which set out basic objectives, and are then followed by subsidiary directives known as daughter directives which set out more precisely the objectives which the Member States must achieve (eg on emission limits and targets etc). For example, EC Directive 76/464 is a framework directive relating to dangerous substances discharged into the aquatic environment of the EC, and it has been followed by seven daughter directives each of which relates to specific dangerous substances. Failure by a Member State to transpose a directive may be the subject of proceedings in the European Court of Justice, instituted by the European Commission; and provisions of a non-transposed directive may none the less be enforceable against a Member State, or an emanation of it, under the doctrine of direct effect *(qv)*. Further reference: House of Lords Committee on the European Community, *Implementation and Enforcement of Environmental Legislation* (Session 1991–92, HL Paper 53).

directive waste

Any substance or object which falls within the categories set out in Part II of Schedule 4 of the Waste Management Licensing Regulations 1994 and which the producer or the person in possession of it discards, or intends or is required to discard. There are certain substances and objects which, if they come within the terms of Article 2 of the EC Framework Directive 75/442 on waste (as amended by Directives 91/156) and 91/692), are excluded from the scope of this definition, eg gaseous wastes. The term 'discard' is an essential feature of this definition, but is not actually defined in Directives 75/442 or 91/156. 'Producer' also has a specific meaning for these purposes, ie 'anyone whose activities produce Directive Waste or who carries out preprocessing, mixing or other operations resulting in a change in its nature or composition' (see reg.1(1) of the 1994 Regulations). It is only directive waste which is subject to the waste management control regime under the Environmental Protection Act 1990, Part II. Waste regulation authorities *(qv)* have different views on what is waste – sometimes according to their regional locations.

director general of electricity supply (OFFER)

See *Office of the Director General of Electricity Supply (OFFER).*

directors' and officers' liability

Secondary liability which arises where the criminal conduct of a company is ascribable to directors or officers of the company, so that they can be personally prosecuted for the offence ostensibly committed by the company. Under the Environmental Protection Act 1990, s.157(1), if an offence under that Act is committed by a company, any director or officer of that company will also be guilty of the offence if it can be proved that the offence was committed with his consent or connivance, or attributable to any neglect by him. Similar provisions are contained in the Water Resources Act 1991, s.217(1), the Food Act 1990 and the Consumer Protection Act 1987, s.37.

discard

See *directive waste.*

discharge consent

A consent granted by the National Rivers Authority (NRA) *(qv)* for discharge of matter into controlled waters *(qv)*. It is a defence to certain water pollution offences under the Water Resources Act 1991, s.85, that the discharge concerned was authorised by discharge consent. The statutory framework relating to discharge consents (eg application, issue, revocation etc) is prescribed by Sched 10 to the Act. Discharge consents apply only to point sources *(qv)*. In deciding whether to consent to a discharge, and upon the conditions to be imposed, the Authority must have regard to any water quality objectives (WQOs) *(qv)* for the receiving waters, and where statutory WQOs have been set, the Authority must set conditions in consents to limit any discharges so that the appropriate classification is met.

Numeric conditions are applied to discharges which have the most potential to harm the environment, and will apply emission limits to individual substances or attributes of substances discharged, generally specified as concentrations and/or load, together with flow. Trade effluents are normally subject to absolute limits, but municipal sewage effluents are allowed to exceed certain limits in a proportion of samples (though this approach will change with full implementation of EC Directive 91/271/EEC on urban waste water treatment). The Authority's policy is to set discharge limits which are individually based on local circumstances, rather than to rely upon uniform emission standards *(qv)*. In relation to processes prescribed for the purposes of integrated pollution control (IPC) *(qv)*, consent to any discharge to controlled waters will form part of the IPC authorisation, though the Authority must be consulted. Further reference: NRA, *Discharge Consents and Compliance: the NRA's Approach to Control of Discharges to Water* (Water Quality Series No 17; 1994).

disposal levy

See *landfill levy.*

diurnal

Daily or recurring every day, such as the diurnal cycle of air pollution concentrations.

drainage

Defined by the Land Drainage Act 1991, s.72(1) as including:

1 defence against water (including sea water);

2 irrigation, other than spray irrigation;

3 warping; and

4 the carrying on, for any purpose, of any other practice which involves management of the level of water in a watercourse.

Item (d) is an addition proposed by the Environment Bill 1995. See also *land drainage*.

drinking water

Potable water. Water used for human consumption. Its quality is controlled by the following EC Directives:

- 75/440 Directive concerning the quality required of surface water intended for the abstraction of drinking water in the Member States
- 79/869 Directive concerning the methods of measurement and frequency and sampling and analysis of surface water intended for the abstraction of drinking water in the Member States
- 80/778 Directive relating to the quality of water intended for human consumption

They have been transposed into UK law by the Water Supply (Water Quality) Regulations 1989 (SI 1989 No 1147) and the Surface Waters (Classification) Regulations 1989 (SI 1989 No 1148). The first Regulations lay down standards for drinking water quality including prescribed concentrations or values of various matter in water. The Schedule to the second Regulations sets out parameters in respect of those surface waters which may be used for the production of drinking water. (See also the selected list of environmental directives on page 255).

Drinking Water Inspectorate

An Inspectorate within the Department of the Environment appointed as the 'assessors' envisaged by the Water Industry Act

1991, s. 86, for monitoring the quality of water supplied by statutory water undertakers *(qv)* in England and Wales, and for initiating enforcement action to secure compliance with the legislation. The Secretary of State is required under s.18 to take enforcement action unless he is satisfied that the contravention is trivial, or that the company has already given an undertaking to take steps to secure compliance. The Inspectorate is responsible for enforcement action when a water quality standard *(qv)* set by the Water Supply (Water Quality) Regulations 1989 (SI 1989 No 1147), reg. 3, is breached; and where there has been a failure to comply with other requirements, such as those relating to sampling, analysis or water treatment. The Inspectorate undertakes a continuous sampling programme, and publishes an annual report.

dry weather flow (DWF)

The sewage *(qv)*, together with infiltration if any, flowing in a sewer *(qv)* in dry weather or the rate of flow of sewage, together with infiltration if any, in a sewer in dry weather.

Duales System Deutschland (DSD)

A German scheme for the recycling of packaging waste, established by German industry in response to the Packaging Ordinance which requires companies that produce and use packaging to take it back, or arrange for it to be taken back. Participating manufacturers use a 'green dot' on their packaging to indicate that it can be recycled, and that a payment has been made by the manufacturer to the scheme. German households have been so diligent that more packaging has been collected than anyone had expected, yet not enough capacity exists to recycle the rubbish into anything people might use, let alone buy. The problem is especially acute for plastics. Germany has enough facilities to cope with 120,000 tonnes at most of the 400,000 tonnes annually to be collected. The requirement that the material should be recycled and not incinerated has caused major market distortions, and material has been transported to other countries, not only in the EC but also as far away as Indonesia and Argentina. Secondary materials systems outside Germany for collecting and recycling have been disrupted, and their viability put at risk (an effect which is particularly acute with plastics and paper). The mixed success of the scheme has stood in the way of the adoption of a proposed EC Directive on packaging and packaging waste. Further reference: House of Lords Select Committee on the European Communities, *Packaging and Packaging Waste* (Session 1992–93; HL Paper 118). See also *plastics recycling*.

due diligence

Reasonable care. It is a statutory defence to environmental liability in some criminal contexts for the defendant to prove that he took all reasonable precautions and exercised due diligence (see eg Environmental Protection Act 1990, s.33(7): offences in relation to the unauthorised deposit, treatment, keeping or disposing of waste; Water Industry Act 1991, s.70(3): water undertaker *(qv)* supplying water unfit for human consumption). An obligation to exercise due diligence may also arise under contract, such as in relation to the sale and purchase of shares in a company. Can also mean research by the purchaser into the standing of a company which is being purchased and, if it is environmental due diligence, this can be difficult to achieve in the customary short time given, especially in relation to such matters as site contamination and levels of compliance with applicable environmental laws.

dumping at sea

The practice of transporting and disposal of wastes in the open sea, usually done beyond coastal or tidal areas and outside the normal territorial waters of any nation. Often used for liquids or sludge which are readily dissolved or diluted by sea water, although also practised for solids such as estuarial dredgings or mineral processing wastes such as colliery spoils. The categories of waste which have in the past been dumped at sea include liquid industrial waste, solid industrial waste (largely fly ash from power stations) radioactive waste, waste munitions from the armed forces, sewage sludge, waste rock from collieries and dredged materials. The greatest waste by volume in the past has been sewage sludge *(qv)* (see further Royal Commission on Environmental Pollution (RCEP) *(qv)* 11th Report, *Managing Waste: The Duty of Care* Cmnd 9675 (1985)). New developments in the control of marine deposits have now been embodied in the Convention for the Protection of the Marine Environment of the North East Atlantic which will replace the 1972 Oslo Convention on the North East Atlantic Sea and North Sea, under which dumping previously was undertaken. When the former is fully implemented the dumping of liquid and solid industrial waste and of waste munitions will have ceased. EC Directive 91/271 on urban waste water treatment (Article 14) commits the United Kingdom to phasing out the dumping of sewage sludge at sea by the end of 1998. The dumping at sea of waste rock from collieries (minestone) will cease by the end of 1997 unless no practicable land-based methods are available. The remaining permitted dumping at sea will be limited, therefore, to dredging ports and harbours and fish wastes from fish-processing vessels. These activities will be regulated by international agreements under the Oslo Convention and its eventual replacement. The amount of dredged materials dumped at sea has fluctuated in recent years but is expected to continue at much the same general levels.

No licences are now issued to dump heavily contaminated dredge spoil. Tighter controls over the load limits and dumping licences, coupled with changes in the nature of the sediment to be dredged, resulting from stricter controls over discharges of effluents into rivers and estuaries and better treatment of sewage, are leading to lower levels of deposit of potentially polluting substances in estuaries and coastal waters. The Control of Pollution (Landed Ships' Waste) Regulations 1987 SI 402 regulate tank washings or garbage landed in Great Britain and provide the rules for reception facilities. They are complementary to the Dangerous Substances in Harbour Areas Regulations 1987 SI 37. Further reference: House of Lords Select Committee on the European Committees, *Dumping of Waste at Sea* (HMSO, 1986); *Sustainable Development: The UK Strategy* (HMSO, 1994).

dust

Fine particles of solids of a size which makes them easily air-borne in turbulent air conditions. Dust from waste may be predominantly of a silica base, but can be chemically very complex, with condensed volatiles and soluble metal salts making it particularly hazardous, poisonous or polluting.

duty of care as respects waste management and transportation

A duty, underpinned by criminal sanctions, imposed by the Environmental Protection Act 1990, s.34, on the holder of controlled waste *(qv)*, to take reasonable steps to ensure against the escape of waste from his or any other person's control; to ensure that on the transfer of the waste it goes only to an authorised person, and with proper documentation; and to prevent any unauthorised disposal, treatment or transfer of it by any person. The 'holder' is an importer, producer, carrier, keeper, disposer, treater or broker of waste. Hence the duty covers those responsible for the arising and importation of waste, and those who subsequently have control over the waste at whatever point in the chain from arising to final disposal.

The duty applies regardless of whether the waste is held in a commercial capacity, but it does not apply to occupiers of domestic premises from which household waste is produced. In determining the reasonableness of any activity to which the duty applies, regard is to be had to the particular capacity in which waste is held. The duty will normally be discharged by the person who holds waste by providing proper facilities for its containment, and by ensuring that appropriate and entirely adequate documentation is made available to an approved person authorised to handle waste on its transfer. Advice on the duty is contained in an official *Code of Practice: Waste*

Management Duty of Care (HMSO, 1991), which is admissible in evidence in any proceedings under these provisions (s.34(10)).

Earth Summit

The United Nations Conference on Environment and Development *(qv)* held in Rio de Janeiro in June 1992. See *Rio Summit.*

eco-audit

See *environmental audit.*

ecoclimate

The local climate of a particular habitat or ecosystem.

ecolabel

A label identifying manufactured products that satisfy certain conditions of environmental significance or criteria. There are various national and international ecolabelling schemes. Originally, ecolabelling in the EC was to be enforced by a regulation but because the Commission of the European Communities found the subject far more complicated than they had anticipated, it will now become a voluntary regulation. Ecolabels are awarded to products on the basis of a full lifecycle analysis, from production to final disposal. Ecolabels are so far in use for soil improvers, and work is in hand for approval of household detergents.

ecological damage

Harm or damage caused to an ecosystem or ecosystems or part of an ecosystem.

ecology

The branch of biology that deals with organisms' relations to one another and to the environment in which they live.

Eco-management and Auditing Scheme (EMAS)

A voluntary European Communities (EC) scheme to encourage industry to undertake positive environmental management including regular audits and to report to the public on its environmental performance. The scheme is the subject of an EC Regulation (1836/93) which came into effect on April 10, 1995, and is closely related to British Standard 7750 on a Specification for Environmental Management Systems *(qv)*. ISO 14001 *(qv)* is also being developed in parallel with it.

EMAS aims to promote continuous improvements in the environmental performance of industrial activities by the establishment and implementation by companies of environmental policies and management systems, the systematic evaluation of their environmental performance, and the provision of information on environmental performance to the public. The scheme is open to companies undertaking an 'industrial activity', as defined. The essential elements of the scheme are set out in Article 3 of the Regulation, and consist of: a company environmental policy, an environmental review, an environmental programme, periodic environmental audits, objectives for continuous improvement, an environmental statement, verification, and registration of site(s).

Economic and Social Committee (ECOSOC)

An institution of the European Communities (EC) established under Articles 193 to 198 of the Treaty of Rome. Members are appointed by the Council of the European Communities and consist of representatives of industry, agriculture, employees, the professions and the general public. Environmental interests are represented. ECOSOC's role is purely advisory though in some instances the European Commission and the Council of the European Communities are obliged to consult with ECOSOC. Its influence looks likely to decline as that of the Parliament increases.

economic instrument

An instrument which attempts to induce desired changes in behaviour by altering price signals rather than by resorting to legal coercion, and which has come to assume some importance in relation to environmental regulation. It is a generic term, which covers a variety of specific instruments, such as fees or charging instruments, which operate through attaching a specific price to a defined quantity of emissions or discharges to the environment; tradable permits, which allow companies to trade permitted emission rights with one another; auctions, in which the state auctions off a limited range of property rights in emissions of pollutants, and which may be tradable or non-tradable; offset, or 'bubble' approaches, in which a facility is permitted to increase emissions in one area, or of one kind, on condition that comparable reductions are made regards other areas, or other emissions; product labelling, as a basis upon which consumers may make informed choices as to possible environmental impacts of any such choice; and subsidies for products or processes that are to be encouraged, that might not otherwise be competitive in market terms.

The European Commission has opened up consultation on the role for economic instruments in the specific context of harmonising industrial competitiveness and environmental protection (Green Paper, *Industrial Competitiveness and Environmental Protection*). The UK Government has gone as far as stating that there will be a presumption in favour of economic instruments in areas that would traditionally have been viewed as the province of legal compulsion (*This Common Inheritance, Second Year Report* (HMSO, 1992)). However, many difficulties remain in designing and administering instruments that will set appropriate price signals, some of which are analysed in *Landfill Costs and Prices* (HMSO, 1993). Problems of this kind have yet to be satisfactorily resolved, and economic instruments remain relatively undeveloped in both a domestic and European Communities environmental policy context. In the UK the lower price for lead-free petrol is an easy example of economic instrument used to attain environmental objectives.

ecosphere

That portion of the Earth which includes the biosphere and all the ecological factors which operate on the living organisms which it contains.

ecosystem

A community of interdependent organisms and the environment which those organisms inhabit, eg ponds and pond life. See also *habitat.*

ecotoxics

Substances or preparations which present or may present immediate or delayed risks for one or more sectors of the environment.

effective chimney height

The lowest chimney height for effective combustion and for dispersal of emissions consistent with environmental requirements.

effluent

Any liquid, including particles of matter and other substances in suspension in the liquid (Water Industry Act 1991, s.219(1)).

elasticity

The property of a body or material of resuming its original form and dimensions when the forces acting upon it are removed. If the forces are sufficiently large for the deformation to cause a break in the molecular structure of the body or material, it loses its elasticity and the elastic limit is said to have been reached.

electromagnetic fields (EMF)

Fields which surround electrical equipment, and whose strength diminishes rapidly with distance from the source. Controversy rages as to the extent of the effect of low levels on humans and other animals, especially in the case of overhead transmission lines. In *R v Secretary of State for Trade and Industry, ex parte Dudderidge* (October 3, 1994) the High Court dismissed an action seeking to require the Secretary of State to make regulations under the Electricity Act 1989 for public protection from EMF. It was accepted that the present state of research was such that no clear relationship could be identified, but the applicants urged the court to give direct effect *(qv)* to the precautionary principle *(qv)* enshrined in the Treaty of Rome, Article 130r. The court declined, holding that the Treaty envisaged that effect would be given to the principle by EC measures, and that it did not constitute an obligation upon Member States themselves to take specific action.

electromagnetic radiation (EMR)

Radiation comprising electric and magnetic energy travelling through a vacuum or a material, eg x-rays, light, infrared radiation, gamma rays, ultraviolet radiation and radiofrequency radiation.

element

A substance which cannot be reduced by chemical means to a simpler substance.

emission source

The point of origin of any emission. See also *diffuse source pollution; non-point source pollution; point source pollution.*

emission standard

A standard limiting the discharge of matter into air, water or land from a particular source. May be imposed as uniform emission standards *(qv)* relating to all emissions of particular substances; or may be translated into specific emission limits imposed in a discharge authorisation. See also *environmental quality standard; process standard; product standard.*

empty

The concept of empty is a question of fact in all the circumstances. In *Durham County Council v Tom Swan & Co Ltd* [1995], Environmental Law Reports, 1972 the defendant held a waste disposal licence which permitted the use of part of its premises for the disposal of a quantity of drums each week. The drums contained small residues of solid phenol waste in a quantity of less than 1% of original volume. The drums were held to be empty.

endangered species

A species in danger of becoming extinct. See also *Convention on International Trade in Endangered Species (CITES).*

endemic species

A species of plant or animal native to, and confined to, a certain country or area.

end of pipe

An approach to pollution control which concentrates upon effluent treatment or filtration prior to discharge into the environment, as opposed to making changes in the processes giving rise to the wastes.

Energy Efficiency Office (EEO)

A division of the Department of the Environment whose objective is to promote greater efficiency in the use of energy, with a view to bringing social and economic benefits and reducing emissions of greenhouse gas *(qv)*. EEO runs a best practice programme which disseminates information on energy performance.

Energy Savings Trust (EST)

A component of the UK Government's programme for reducing the emission of greenhouse gases. The EST was established in November 1992 by the Government, British Gas and the electricity supply companies, with the objective of providing financial incentives for the more efficient use of energy, particulaly by domestic households. It provides subsidies for the installation of high-efficiency gas condensing boilers, for low-energy light bulbs and for combined heat and power *(qv)* schemes, It has a target of savings of at least 2.5 MtC by the year 2000.

Energy Technology Support Unit (ETSU)

A unit of the Department of Trade and Industry (DTI) which provides advice on potential energy savings on plant and equipment.

enforcement notice

A notice served by a regulatory agency requiring that a specified breach of regulatory control should be rectified. This is usually instead of, but may be in addition to, any criminal proceedings for the same breach. Its purpose is to remedy the breach rather than to punish its occurrence. The normal statutory requirements are that:

1 the notice must be served on the person (or one of the persons) responsible for taking remedial action,

2 it must state what matters are considered by the regulator to be a breach of relevant controls, specifying these,

3 it must specify with sufficient particularity what steps must be taken, and

4 it must stipulate a date when it is to take effect (often at least 28 days from service), and by when the action must be completed, giving a reasonable time for this.

Failure to comply with the notice is itself a criminal offence, and there is a right of appeal against the notice to the Secretary of State. Examples of this procedure include the Town and Country Planning Act 1990, s.172 (breach of planning control); Environmental Protection Act 1990, s.13 (breaches relating to integrated pollution control and air pollution control); Radioactive Substances Act 1993, s.21 (breaches relating to conditions on authorisiation); and the Planning (Hazardous Substances) Act 1990, ss.24, 24A and 25 (breach of hazardous substances control).

English Heritage (EH)
The working name adopted by the Historic Buildings and Monuments Commission, which was established by the National Heritage Act 1983. Its functions include giving advice in relation to listed buildings, conservation areas and archaeology situated in England including advising the Secretary of State for National Heritage on the listing of historic buildings and the scheduling of ancient monuments. It may make grants and loans in relation to historic buildings, land and gardens and for archaeological investigation. It also maintains the registers of historic parks and gardens and battlefields. It is responsible for the management of some 400 sites and monuments, and advises the Secretary of State for the Environment, the Secretary of State for the National Heritage and local authorities, in particular in relation to applications for listed building consent. [*Address: Fortress House, 23 Savile Row, London W1X 1AB; T: 0171 973 3000; F: 0171 973 3001.*]

English Nature
The working name adopted by the Nature Conservancy Council for England, established by Part VII of the Environmental Protection Act 1990 (EPA 1990) to be one of the three statutory bodies succeeding to the previous Nature Conservancy Council (the others being the Nature Conservancy Council for Scotland and the Countryside Council for Wales *(qv)*). It advises the Government on nature conservation in England. It promotes, both directly and through

others, the conservation in England of its wildlife and natural features. It selects, establishes and manages national nature reserves (NNRs) in England and identifies and notifies sites of special scientific interest (SSSIs) *(qv)*. It provides advice and information about nature conservation, and supports and conducts research relevant to these functions. Issues concerning the United Kingdom as a whole, and international issues, are dealt with by the English, Scottish and Welsh bodies acting jointly through the Joint Nature Conservation Committee, which includes representatives from Northern Ireland. [*Address: Northminster House, Peterborough PE1 1UA; T: 01733 340345; F 01733 68834.*]

envelope authorisation

A type of authorisation which confers flexibility on the permit-holder by specifying permitted processes and emissions in terms sufficiently general to allow increases in emissions from one part of the process to be offset against reductions elsewhere. In particular, it is used in respect of chemical processes subject to integrated pollution control (IPC) *(qv)*, to cater for batch manufacture of a variety of different products on multi-purpose plant. Instead of relating to one specific process consisting of precisely defined steps, an envelope authorisation will typically define the particular area of process chemistry to be carried out, list the compounds within this classification that are authorised to be manufactured and define the production and pollution abatement systems to be used together with the release levels expected. See also *tradable permits.*

environment

In Albert Einstein's epigrammatic definition, 'The environment is everything which isn't me'. The Environmental Protection Act 1990 (EPA 1990) makes heavier work of it, providing three alternative definitions for the purposes of different provisions in the Act (ss.1(2), 29(2) and s.107(2)). It extends in all cases to the three media of air, water (including groundwater) and land; for the purposes of the air pollution and integrated pollution controls of Part I EPA 1990, it is expressly stated to include air within buildings and other natural or man-made structures above or below ground.

Man and other living organisms supported by these media are not included in these definitions of the 'environment', though in common parlance, the term is often used in this broader sense, particularly in the context of environmental protection. In addition, the elision of health and safety practice and law with environmental control is not sufficiently recognised in the UK in the late 1990s. For example, most consignment control regulations are health and safety based, but risks attendant on transport practices are generally conceived to be environmental matters.

Environment Agency

The body to be established by the Environment Bill 1995 to carry out in England and Wales the functions of the National Rivers Authority, Her Majesty's Inspectorate of Pollution and the Waste Regulation Authorities, as well as certain functions previously undertaken by the Secretary of State for the Environment, (eg responsibilities for 'special category effluent' under the Water Industry Act 1991, and controls over the use of sewage sludge in agriculture). These functions will include responsibilities for land drainage and coastal defence. The corresponding body for Scotland (with additional responsibilities for air pollution control), is the Scottish Environment Protection Agency. See also *European Environment Agency*, and (US) *Environment Protection Agency*.

Environmental Action Programme (EAP)

Successive programmes adopted by the European Commission. EAPs were adopted in 1972, 1977, 1982, 1987 and 1993. The most recent, the 5th Environmental Action Programme, *Towards Sustainability*, sets out the Commission's proposals for the period 1993–2000. The programme is not a legislative instrument, but a policy statement setting out long-term objectives and performance targets. Further reference: House of Lords Select Committee on the European Communities, *5th Environmental Action Programme: Integration of Community Policies* (Session 1992–93; HL Paper 27). See also *playing field, level*.

environmental appraisal

A process of evaluation of all the different options, including the environmental and other costs and benefits, before taking a decision on an initiative or project which may affect the environment.

environmental assessment

See *environmental impact assessment*.

environmental audit

A systematic review or appraisal of the environmental performance and status of an organisation or its processes, sometimes known as an eco-audit. Environmental audits are commonly undertaken as an:

1 internal examination conducted by a company in respect of its own operations to assess its environmental compliance and performance. For example, the European Communities

(EC), Eco-management and Auditing Scheme (EMAS) *(qv)* and British Standard 7750 on a specification for environmental management systems *(qv)* require the conduct of environmental audits to assess performance under an environmental management system;

2 external examination of a company conducted to ascertain and quantify that company's actual and/or potential environmental liabilities. This kind of audit will normally be conducted as part of a due diligence *(qv)* exercise by a purchaser, underwriter or lender in a wide range of commercial transactions, eg share purchase. The investigation will normally be undertaken in three phases. Phase I is primarily a desk-based site evaluation, Phase II involves intrusive work to determine the extent of problems revealed during Phase I and Phase III addresses the options for dealing with the problems found to exist by remediation;

3 associate audit conducted by a company with respect to the environmental performance and compliance of its suppliers, distributors, agents or its licensees. The aim of an associate audit is to ensure that associated organisations are environmentally sound and have standards of environmental performance which correspond to those of the company itself.

A phased approach to auditing is generally as follows:

- *Phase I audit*: an initial, visual review of an operation, its activities and associated paperwork. An initial site visit is usually undertaken, with discussions with key staff. A Phase I audit usually covers all aspects of potential environmental impact from the current site and its previous operations. These results are considered in the context of the environment of the area, in particular the local importance of ground and surface waters and the nature of the underlying geology, in order to assess the relative significance of any impact. This phase highlights any particular areas of concern where further, more involved, research may be prudent.

- *Phase II audit*: a more detailed review and site inspection of the particular areas of concern identified from the Phase I audit. This stage normally involves the sampling and analysis of materials or emissions. The Phase II audit will detail (and should always quantify) any potential liabilities and any clean-up or process control requirements.

- *Phase III audit*: involves the design, commission and supervision of any clean-up or process control requirements which are required or prudent on the basis of the Phase II audit.

Environmental Auditors Registration Association (EARA)

A United Kingdom organisation established in 1992 to develop and promote standards for environmental auditors, and in particular to set up a register of auditors with prescribed degrees of competence. Depending on their qualifications, training and experience, individuals may be entered on the register as Associate Auditors, Environmental Auditors or as Principal Auditors. [*Address: The Old School, Fen Road, East Kirkby, Lincolnshire, PE23 4DB; T: 01790 763613; F: 01790 763630.*]

environmental consultants

Individuals who work for themselves or as part of larger organisations and who provide advice on environmental matters. Most are professionals in one or more technical fields relevant to, for example, the investigation and remediation of potentially polluted land, water and air, with the qualifications of engineering, geology, chemistry, hydrogeology, landscaping, environmental economics, etc. The Association of Environmental Consultancies is a trade association of organisations that have expertise in this area.

environmental health officer (EHO)

Local government official responsible for the management and enforcement of a variety of environmental controls. Functions relate to air pollution control (under Part I of the Environmental Protection Act 1990), certain statutory nuisances, public health and food and hygiene in shops and restaurants.

environmental impact assessment (EIA)

A technique and a process for gathering information on the likely environmental effects of a proposed development or activity; this should properly be carried out while the project is still at the design stage, so that any adverse effects can be addressed and, where practicable, minimised in the plans. The approach was pioneered in the United States, where it was made mandatory for major construction projects by, or financed by, the Federal Government. It was subsequently taken up by the EC and is the subject of Directive 85/337. Implementation of the Directive in the United Kingdom is principally by the Town and Country Planning (Assessment of Environmental Effects) Regulations 1988 (SI 1988 No 1199) as amended, for England and Wales, the Environmental Assessment (Scotland) Regulations 1988 (SI 1988 No 1221) for Scotland, and the Planning (Assessment of Environmental Effects) Regulations (Northern Ireland) (SR No 20). There are numerous other related Regulations, mostly concerned with categories of projects covered by the Directive but which are not subject to the standard planning consent procedures in the United Kingdom.

The process is made mandatory for major projects, being those listed in Annex I to the Directive and in Schedule 1 to the 1988 Regulations. For a further series of projects, listed in Annex II to the Directive and in Schedule 2 to the 1988 Regulations, an environmental statement may be required by the planning authority (subject to appeal) if it is considered that the project in question 'would be likely to have significant effects on the environment by virtue of factors such as its nature, size or location'. The assessment procedure essentially involves the developer submitting an environmental statement *(qv)* to the relevant authority. Copies also go to certain statutory consultees. The authority may seek further information, and must lay the statement open for comment by the public before determining any application for planning consent for the project. A developer may volunteer an environmental statement even though it cannot be legally demanded, and this may have procedural and other advantages for him. Formal guidance on the procedures is in Department of the Environment Circular 15/88 or Welsh Office Circular 23/88. Further guidance is in a Department of the Environment explanatory booklet *Environmental Assessment – A Guide to Procedures* (HMSO). An international regime for environmental impact assessment for projects likely to have significant adverse transboundary effects has recently been agreed, but still awaits implementation. See also *Convention on Environmental Impact Assessment in a Transboundary Context.*.

environmental impairment liability (EIL) insurance
Insurance cover which extends not only to accidental damage, but also to environmental impairment as a result of, for example, gradual leakage from underground tanks or waste disposal practices. Current uncertainties about future liabilities for environmental damage (see also *contaminated land* and *environmental liability*) have made insurers wary about offering EIL cover.

environmental information and observation network (EIONET)
A network being established by the European Environmental Agency

(qv), which will co-ordinate work and forge links between some 400 existing research and policy institutes in Member States.

environmental liability

Not a technical term, but used to describe liability both under criminal law and under civil law. Criminal liability *(qv)* in the environmental context typically arises from prosecution by a regulatory authority for non-compliance with environmental legislation, and its function is to underpin regulatory requirements by punishing those who fail to comply with them. Civil liability is a means for charging back the costs of environmental damage to those who cause that damage, and of preventing threatened damage. The limited effectiveness of traditional civil law systems in achieving that objective has led to consideration of how they might be enhanced, hence the European Commission's 1993 *Green Paper on Remedying Environmental Damage* (7092/93 ENV 170 COM(93) 47 final), upon which further lengthy research and consultation is being conducted with a view to the preparation of a draft directive.

environmental lien

An expression used in the US to describe a charge, security or other incumbrance upon title to a property (normally land) to secure the payment of a financial contribution arising out of clean-up or other remediation of hazardous substances or petroleum products. It includes, but is not limited to, liens imposed pursuant to Comprehensive Environmental Response, Compensation and Liability Act (CERCLA) *(qv)* and similar state or local laws. Such a lien may be given priority over other incumbrances on title, such as the mortgage security of a lender.

environmentally sensitive area (ESA)

Area designated under the Agriculture Act 1986, s.18 in which the Minister of Agriculture, Fisheries and Food considers it to be particularly desirable to conserve, protect or enhance environmental features by the maintenance or adoption of particular agricultural methods. The Act enables the Minister to enter into a management agreement with any person having an interest in agricultural land in an ESA under which that person agrees (in consideration of payments to be made by the Minister) to manage the land in accordance with the agreement. The relevant Order designating each ESA specifies those requirements which *must* be and those which *may* be contained in an agreement relating to any part of that area, including *inter alia* requirements as to agricultural practices, methods and operations, the installation or use of particular equipment, public access and conservation plans.

environmental management system

A system designed to enable the management of an organisation to assess and control their environmental performance. Numerous different systems have been developed, but they generally have in common an iterative process consisting of: defining an environmental policy; reviewing the organisation's activities against the policy objectives; establishing a programme of action in respect of activities which do or may (eg in the event of an accident) impinge on the environment so as to manage these activities and to measure and control their impacts; periodic audits of these activities; setting targets for improvement and revising the environmental policy as appropriate, and repeating this cycle.

The British Standard 7750 on a Specification for Environmental Management Systems (BS 7750) *(qv)* sets out a (voluntary) quality standard for environmental management systems, which is very similar to that required for registration under the Eco-management and Auditing Scheme (EMAS) *(qv)* and to the proposed ISO 14001 *(qv)*.

Environmental Protection Act 1990 (EPA 1990)

One of the principal environmental statutes in the United Kingdom, covering integrated pollution control and local authority air pollution control (Part I), waste (Part II), statutory nuisances (Part III), litter (Part IV), some aspects (mostly deliberate release) of genetically modified organisms (Part VI), and the establishment of separate nature conservation bodies for England, Wales and Scotland. The Act also provides various powers over miscellaneous matters (eg hazardous substances and articles, control of dogs, and burning of straw and stubble).

A new Part IIA is proposed by the Environment Bill 1995 and will deal with contaminated land *(qv)*.

Environmental Protection Agency (EPA)

(Also sometimes referred to as the USEPA.) The Federal Agency with responsibility for environmental issues in the United States of America. Several individual states also have their own environmental protection agencies.

environmental quality objective (EQO)

A statement of the quality to be aimed for in a particular aspect of the environment, and which can then be used as a basis for programmes to secure the necessary improvements to bring it about, and for devising appropriate emission limits to be imposed upon authorisations to discharge matter into that environment. The objective will usually be framed in qualitative terms, but then converted into a set of environmental quality standards *(qv)* specifying the maximum admissible concentrations of particular contaminants, and the minimum permissible level of desirable features. EQOs have been used for many years in the United Kingdom in relation to river quality, but on a non-statutory basis and relying upon the classification process devised by the 1912 Royal Commission.

They have now been placed on a statutory basis under the Water Resources Act 1991, ss.82 – 84 which enables the Secretary of State to prescribe a system for the classification of the quality of different descriptions of controlled waters *(qv)*, then to require the National Rivers Authority to establish water quality objectives for them. It then becomes the duty of the Secretary of State and the NRA to exercise their water pollution powers so as to ensure that the water quality objectives are secured at all times. EQOs have also been used on a statutory basis in relation to air pollution. See also *air quality standard; environmental quality standard; water quality objective; water quality standard.*

environmental quality standard (EQS)

A standard, normally expressed in quantitative terms, which specifies the maximum permissible level of a contaminant in a specified environmental medium and at a particular location (eg urban area, or stretch of a river), or specifies the minimum permissible level of some positive environmental attribute (eg oxygen). It relates to the receiving environment, unlike an emission standard *(qv)*, which relates to the content of an emission to the environment; but it may then be used as a basis for setting emission standards for discharges into that environment so as to ensure that they will not cause the EQS to be breached, and hence advance the environmental quality objective for that environment. See also *air quality standard; water quality standard.*

environmental reporting

The reporting by organisations on their environmental performance, typically to shareholders in the annual report, but also by way of separate reports for other groups such as employees and the general public. It is an essential requirement for registration under the Eco-management and Auditing Scheme (EMAS) *(qv)*, and also under the Valdez Principles *(qv)*. Guidance on the content of reports and the methodology used to prepare them has been produced by the World Industry Council for the Environment and the Public Environmental Reporting Initiative.

environmental statement

A document(s) that must be submitted by the proposer of a project as part of the process of environmental impact assessment *(qv)* of the project. For the United Kingdom, the requirements are specified in Annex III to EC Directive 85/337, transposed into national law by Schedule 3 to the Town and Country Planning (Assessment of Environmental Effects) Regulations 1988 (SI 1988 No 1199) (and see also the other regulations listed under *environmental impact assessment*).Guidance on the content of an environmental statement is given in *Environmental Assessment – A Guide to the Procedures* (HMSO) at paragraphs 24 to 26 and Appendix 3.

Environmental Technology Best Practice Programme

A programme jointly launched by the Department of the Environment and the Department of Trade and Industry in January 1994 which aims to promote cost-effective waste minimisation strategies and cleaner production techniques within industry. The first three target areas are volatile organic compounds (VOCs) *(qv)*, foundries, and metal finishing and surface engineering. [*Environmental Hotline: 0800 585794.*]

environmental trusts

The concept of environmental trusts was first floated by the Government in March 1995 in its Consultation Paper on the landfill levy *(qv)*, with the purpose of repairing the damage from old landfills and as a substitute for legislation enacting retroactive penalties on past polluters.

The Government proposes the trusts be funded from a rebate out of the proposed landfill levy increased by a contribution from the waste disposal industry. The latter have, however, claimed that they are being penalised twice; by paying more in the form of the landfill levy because it is levied on a flat rate on gate charges and also by having to pay for other public or private sector operators' past mistakes.

The trusts will be independent of the Government and aim to fund research into sustainable waste management practices as well as restoration of closed landfills, particularly where the liability is unclear.

Environment Bill 1995

A Bill introduced to the United Kingdom Parliament in November 1994 which is expected to be enacted in 1995. It proposes the creation of a new Environment Agency *(qv)* for England and Wales, and a new Scottish Environment Protection Agency, in order further to strengthen the effectiveness of integrated pollution control (IPC) *(qv)*, to reduce the number of regulators industry will have to deal with and to provide a strong independent voice to influence the adoption of better environmental standards and practices. It also makes fresh provision in respect of the remediation of contaminated land and for liability for pollution from abandoned mines; and it provides a framework for establishing new authorities for national parks *(qv)*.

Environment Protection Advisory Committee

A committee to be established and maintained by the Environment Agency *(qv)* for each of the different regions of England and Wales to advise it on the manner in which the Agency carries out its functions in that region. The Environment Bill 1995 proposes to impose a duty on the Agency to establish such committees.

Environment Technology Innovation Scheme (ETIS)

A scheme launched in October 1990 and jointly managed by the Department of the Environment and the Department of Trade and Industry, with the aim of supporting research leading to new techniques, processes, materials or equipment, mainly in four broad categories: cleaner technologies, recycling, waste or effluent treatment/disposal and environmental monitoring. There was an initial three-year term for the submission of applications for research support.

enzyme

A biological catalyst which increases the reactivity of a specific substance or group of substances. This group of substances is known as a substraite. Enzymes are vital for many reactions which take place in a living organism (eg the hydrolysis of fats, sugars and proteins). Enzymes are susceptible, however, to a wide variety of substances which act generally as poisons; they may be destroyed and inactivated by, for example, insecticides.

epidemiology

The study of environmental, personal and other factors that determine the incidence of disease, often involving a study of persons in their working or living environment to assess any correlation with any disease which they may have suffered; for example, any linkage between an increase in the incidence of infantile asthma and motor vehicle emissions; or between ingestion of nitrates and 'blue baby' syndrome.

erosion

The effect of wind and water on soil and substrata, leading to loss of fertile soil, and in extreme cases, desertification, and to the collapse of coastal cliffs.

Espoo Convention

See *Convention on Environmental Impact Assessment in a Transboundary Context.*

ethical investment

Investment in which the performance of companies is measured against ethical and/or environmental criteria as well as purely financial ones. What those criteria may be and how stringently they are applied to every investment is a matter for the investing institution – there are currently no generally accepted rules.

European Commission

The central bureaucracy of the European Communities. Made up of 17 members (called Commissioners) appointed by the governments of the 15 Member States and 'chosen on the grounds of their general competence and whose independence is beyond doubt' (Treaty of Rome, Article 157) and headed by a President appointed from among the Commissioners. Its functions include:

1 proposing Communities policy and legislation;

2 implementing decisions taken by the Council of Ministers and supervising the day-to-day running of Communities policies;

3 enforcing observance of EC law by Member States; and

4 exercising its own powers under the Treaty, notably the enforcement of competition policy. The Commission comprises 27 departments known as Directorates General. DG XI has responsibility for environmental policy. It has initiated five successive environmental action programmes since 1972, and bears primary responsibility for promoting their implementation. [*Address: (UK) Offices of the European Commission, 8 Storey's Gate, London, SW1P 3AT; T: 0171 973 1992; Directorate General XI, Commission of the European Communities, Rue de la Loi 200, B–1048 Brussels, Belgium ;T: 010 32 2 299 1111.*]

European Communities (EC)

The European Economic Community, the European Coal and Steel Community and Euratom (the European Atomic Energy

Community). They were established by different treaties, but their originally separate administrative organs have been merged, so that the European Commission *(qv)* and the Council of the European Communities *(qv)* are now responsible for all three. Following the Treaty of Maastricht, the European Communities have been subsumed into the European Union, which has a somewhat wider jurisdiction, and includes, in particular, a common foreign and security policy, but they remain in existence each with their distinct legal basis.

European Court of Justice (ECJ)

The judicial organ of the European Communities, established by and operating under Articles 4 and 164–188 of the Treaty of Rome, and in accordance with a Statute contained in a Protocol (No B) to the Treaty. Its principal functions are to give preliminary rulings on questions of law referred to it by national courts under Article 177; to determine, at the instance of the Commission or of any Member State acting under Article 169, 170 or 173, the validity of acts of the European institutions or of any other Member State; and to hear appeals brought under Article 173 from decisions of the European Commission. Previously the Court's powers in relation to complaints of failure by Member States properly to transpose directives *(qv)* were largely declaratory, so that it was a somewhat impotent organ for enforcement of European environmental law. However, the Maastricht *(qv)* amendments amended Article 171 of the Treaty to empower the Court to impose a lump sum penalty on a Member State which fails to comply with its judgement. The function of the ECJ when hearing questions referred to it under Article 177 by national courts is to determine the relevant question of law. How the law as thus defined applies to the facts of the particular case that gave rise to the original question is always left to the national court. There is a Court of First Instance attached to the ECJ with jurisdiction to hear appeals from decisions of the Commission in competition and staff cases, and certain matters arising under the European Coal and Steel Treaty. [*Address: European Court of Justice, L–2920 Luxembourg; T: 010 352 4 3031.*]

European Economic Area

An area established in 1992, by an agreement entered into under the Treaty of Rome, Article 238, comprising the European Communities (EC), Austria, Finland, Iceland, Liechtenstein, Norway, Sweden and Switzerland. The objective is to abolish economic frontiers so as to create an internal market between the EC and the other states. Since the conclusion of the agreement, Sweden, Finland and Austria have become full Members of the EC.

European Economic Community (EEC)

The Community established by the Treaty of Rome of March 25, 1957. See also *European Communities (EC).*

European Environment Agency (EEA)

The Agency created by EC Regulation 1210/90, although disputes between Member States delayed its initial establishment until October 1993. Its functions are the collection and dissemination of information relating to the quality of the environment, the pressures on the environment and its sensitivity to those pressures. It is required to give priority to: air quality and atmospheric emissions; water quality, pollutants and water resources; the state of the soil, of the fauna and flora, and of biotopes; land use and natural resources; waste management; noise emissions; chemical substances which are hazardous for the environment; and coastal protection. It is also required to establish and monitor an environmental information and observation network (EIONET) *(qv)*. Although the Agency currently has no regulatory or enforcement functions, Regulation 1210/90 provides for a review of its tasks by October 1995, and its role may then be expanded.

The role of the Agency has been questioned by the House of Lords Select Committee on the European Communities, particularly as to whether the Agency was under any obligation to provide any information, that there were no powers to order data and that the EEA has been set up to operate on a co-operative basis, with no clear definition of its enforcement role. The Department of the Environment's evidence to the Committee stressed that it would like to see the Agency giving more information on the scientific basis on which parameters and environmental Directives were decided. [*Address: Kongens Nytorv 6, DK–1050 Copenhagen K, Denmark.*]

European Environmental Bureau (EEB)

A non-profit making federation of some 150 environmental organisations from both within and outside the European Communities, of which 24 are from the United Kingdom (including eg the Civic Trust, the National Trust, the Town and Country Planning Association, the Council for the Protection of Rural England (CPRE), the Royal Society for the Protection of Birds (RSPB), the World Wide Fund for Nature (WWF-United Kingdom), and Friends of the Earth). It liaises closely with the European Communities (EC) institutions, by way of formal meetings on a regular basis, and more frequent informal ones between the secretariat and officials of the European Commission. Its Board normally consists of one member for each country in the EC.

[*Address: 26 rue de la Victoire, B–1060 Brussels, Belgium; T: 010 322 539 0037; F: 010 322 539 0921.*]

European Inventory of Existing Commercial Chemical Substances (EINECS)
An inventory of some 100,000 chemical substances prepared by the European Commission pursuant to the 6th Amendment (Directive 79/831) to the Classification, Packaging and Labelling Directive 67/548. Its object was to list all such substances that had been on the market in the European Communities (EC) by September 18, 1981 (the implementation date for the 6th Amendment). It was drawn up in accordance with rules laid down in Commission Decision 81/437, which required the Commission to prepare a 'European Core Inventory' and to supplement this with additional substances notified to it that either had been placed on the EC market 'for genuine commercial purposes' between January 1, 1971 and September 18, 1981, or which were monomers from which polymers marketed in the EC in the same period had been made. The inventory is thus a closed list; substances that have been marketed since September 18, 1981 (or which ought to have been notified in the preparation of EINECS but were not) are listed in the European List of Notified Chemical Substances *(qv)*.

European List of Notified Chemical Substances (ELINCS)
A list of chemical substances prepared pursuant to the 6th Amendment (Directive 79/831) to the Classification, Packaging and Labelling Directive 67/548, and consisting of those not contained in European Inventory of Existing Commercial Chemical Substances *(qv)* that have been notified to national authorities by persons wishing to manufacture them in or to import them into the EC after September 18, 1981. The list is prepared by the European Commission in accordance with Decision 85/71, from information supplied by the national authorities, and it issues a revised version of ELINCS annually. The notification procedure in the United Kingdom is governed by the Notification of New Substances Regulations 1993.

European Parliament
The central elected Parliament of the European Communities (EC), under the Treaty of Rome, Articles 137–144. Its members are directly elected from the 15 Member States of the EC, and it is now one of the largest assemblies in the world with 626 members. The Maastricht *(qv)* amendments developed the powers of the Parliament, conferring on it the opportunity to veto legislation approved by the Council of Ministers but still falling short of providing a form of general democratic control over the other institutions. [*Address: European Parliament, Information Offices, 2 Queen Anne's Gate, SW1H*

9AA; T: 0171 222 0411; European Parliament, General Secretariat, L–2929 Luxembourg; T: 010 352 4 3001].

European Recovery and Recycling Association (ERRA)

An organisation concerned with promoting the recovery and recycling of packaging as part of total waste management systems. [*Address: Avenue E. Mounier 83, Bte 14, B–1200 Brussels, Belgium; T: (322) 772 52 52; F: (322) 772 54 19.*]

European Site

A site forming part of the ecological network known as 'Natura 2000' *(qv)*. Such sites are those listed in the Conservation (Natural Habitats, &c.) Regulations 1994 (SI 1994 No 2176) reg. 10, namely 'special areas of conservation' as designated under the Habitats Directive 92/43; 'sites of Community importance' that have been placed on the list referred to in the 3rd sub-paragraph of Article 4(2) of that Directive; a 'site hosting a priority natural habitat type or priority species' in respect of which consultation has been initiated under Article 5(1) of that Directive, either during the consultation period or pending a decision of the Council; and a 'special protection area' classified pursuant to Article 4(1) or (2) of the Birds Directive 79/409 *(qv)*.

European Union (EU)

The union of Member States which are parties to the European Community Treaties and the Treaty on European Union (the Maastricht Treaty) when they co-operate and take action regarding matters of foreign and security policy, defence and the framing of common policies on Justice and Home Affairs; it is not a separate legal entity but has come to be used generally in place of European Communities (EC) (although, for the sake of consistency, this book uses European Community (EC) throughout). Since January 1, 1995 the Union has had 15 Members, upon the accession of Sweden, Finland and Austria. The EC population increased from 349 million to 371 million (which is one and a half times that of the US and 3 times the population of Japan). The EC occupies 1·2 million square miles.

The European Union flag (known as the Union Jacques) remains unchanged with 12 stars (called mullets); 12 it always has been and 12 it always will be, regardless of how many Members join, since the mullets represent absolutely nothing. The flag was introduced as the official emblem of the European Community (EC) as it was then known, by Jacques Delors, the Commission President, in 1986. It dates back, however, to 1955 when it was first adopted by the Strasbourg-based Council of Europe. When Mr Delors borrowed the flag for the EC, it was pure coincidence that the number of stars equalled the number of Members.

European Waste Catalogue (EWC)

The Waste Framework Directive 75/442, as amended by 91/156, required the European Commission to draw up a list of wastes belonging to the categories listed in Annex I to the same Directive in order to harmonise nomenclature within the European Communities; the Commission's list of such wastes, 94/3, is commonly referred to as the European Waste Catalogue.

eutrophication

The process by which a plentiful nutrient supply in a body of water (such as a pond or lake) results in an over-abundance of plant matter and consequential de-oxygenation of the water. Depletion of the oxygen content of the body of water is caused by the bacterial decomposition of dead plant matter, and may lead to the death of aquatic animals dependent on oxygen in the water. Eutrophication can result from the introduction to the body of water of plant nutrients contained in chemical fertilizers, sewage effluent or animal wastes, all of which are sources of plant nutrients such as nitrates and phosphates, and it is sometimes associated with soil erosion and run-off from agricultural areas. It has a specific definition in EC Directive 91/271/EEC on urban waste water treatment, as meaning 'the enrichment of water by nutrients, especially compounds of nitrogen and/or phosphorus, causing an accelerated growth of algae and higher forms of plant life to produce an undesirable disturbance to the balance of organisms present in the water and to the quality of the water concerned' (Article 2(11)).

evaporation

The process by which liquid is converted to a vapour by the application of energy. Evaporation can take place from a wet soil surface, from snow, ice and open water bodies and from vegetation wetted by rain. In addition, vegetation conveys water from the soil to the atmosphere by evaporation mainly through the stomata in the leaves – a process known as transpiration *(qv)*.

exposure limits

The limit of exposure to outside influences beyond which a person or substance will be adversely affected or changed.

Exxon Valdez

The ship that ran aground on Bligh Reef in Prince William Sound, Alaska on March 24, 1989, and discharged 11 million gallons of crude oil. The captain had been drinking, had left the command post as the ship made its way through Valdez Narrows and had left the controls in the hands of unqualified seamen, without telling them that he had activated the ship's 'load program up' device, which steadily increased its speed until it reached ocean-going rate. Exxon spent \$2·5bn on cleaning up the damage. Although the vessel was owned and operated by a fully owned subsidiary of Exxon, criminal charges were brought successfully against the main company, and were settled in 1991 when Exxon agreed to pay \$1bn in damages to the Federal and State Governments over 10 years. In civil proceedings brought by 14,000 fishermen and others whose livelihood had been affected by the spill, a jury in 1994 awarded \$5bn in punitive damages. That award is currently on appeal.

fallout

Commonly used in respect of nuclear fallout where it refers to the deposition of radioactive particles arising from a nuclear explosion. Also refers to deposition of particulates, eg from combustion. Particle deposition is determined by particle size, large particles settling quickly, fine particles remaining in suspension until removed by wash-out or rain-out.

fauna

The animals of a region or epoch. Derived from the name of the sister of the Roman god of animals.

feral

The description of animals or plants that are no longer domesticated or cultivated and have reverted to a wild state.

fertiliser

A processed chemical compound or manure usually containing high levels of nitrates and phosphates, added to soil to raise nutrient levels often with the aim of improving the productivity of land. Run-off of fertilisers into a body of water may result in eutrophication *(qv)*.

final settlement tank

A tank through which the effluent *(qv)* from a percolating filter, or aeration *(qv)* tank, flows for the purpose of separating settleable solids. The former is often called a humus tank.

fit and proper person

Any individual, partnership or body corporate holding a waste management licence *(qv)* issued under the Environmental Protection Act 1990 (EPA 1990) must be considered by the waste regulation authority (WRA) to be a fit and proper person. WRAs may refuse to grant or transfer licences to, or may revoke or suspend the licences of, persons who in their opinion are not fit and proper. The expression is defined in s.74, and guidance provided in *Waste Management Paper No 4* (3rd Edition 1994). In deciding whether a person is a fit and proper person to hold a licence the WRA will have regard to the activities which are to be authorised by the waste management licence and the likelihood that its requirements will be fulfilled. A person shall be treated as not being fit and proper if:

1 it appears to the WRA that the person (or another relevant person) has been convicted of a relevant offence ('another relevant person' for these purposes refers to a director, manager, secretary or other similar officer of a body corporate) ('relevant offences' are principally waste-related offences, and WRAs will take into account not only whether there has been a conviction but also the number of relevant offences committed and their nature and gravity);

2 the management of the activities which are or will be authorised by the licence are not or will not be overseen by a technically competent person; or

3 the WRA considers that the person has not made financial provision adequate to discharge the obligations under the licence and either has no intention of making or is not in a position to make such financial provision.

fixed matrix

A method of forecasting used in assessing the need for, and capacity of, infrastructure which assumes that no changes will occur in the pattern of usage as a result of the provision of a new element. A classic example of the weakness of such methodology is the failure of the Department of Transport to include in their traffic forecasts for the M25 an allowance for additional vehicle journeys undertaken as a result of its provision.

flame ionisation detector (FID)

An analytical device consisting of two electrodes held in a hydrogen flame and connected to an electrometer. It forms part of an apparatus used for gas chromatography analysis, and provides quantitative and qualitative information on the chemical components of samples passing through the apparatus by detecting charged particles which are formed when chemical components separated in the apparatus are ionised in the hydrogen flame. Most compounds containing carbon and hydrogen are detected using this method but it is relatively insensitive to carbon dioxide (CO_2), carbon monoxide (CO), water (H_2O), oxygen (O_2), nitrogen (N_2), ammonia (NH_3), carbon disulphide (CS_2), inert gases and oxides of nitrogen and sulphur. It is particularly suited for analysis of trace organic compounds in air and water samples.

flood defence

The construction, improvement and maintenance of defences against flooding from rivers and the sea. For the purposes of Part IV of the Water Resources Act 1991, it is defined as 'the drainage of land and the provision of flood warning systems' (s.113). Flood defence policy is determined by the Ministry of Agriculture, Fisheries and Food (in Wales, the Welsh Office). The National Rivers Authority is required to exercise a general supervision over all matters relating to flood defence, but to arrange for all its functions in this respect to be carried out by regional flood defence committees, which comprise members appointed by the Ministry of Agriculture, Fisheries and Food (in Wales, the Welsh Office), local authorities and the National Rivers Authority. The National Rivers Authority may give directions to the committees as to how to carry out their functions so far as they impinge upon the Authority's other water management functions (Water Resources Act 1991, s.106).

flora

The plants of a particular region or epoch. Derived from the name of the Roman goddess of flowers.

flue gas desulphurisation

The means by which oxides of sulphur, mainly sulphur dioxide (SO_2), are removed from flue gases. One of the more efficient and widely used methods is to absorb sulphur dioxide into an aqueous solution of potassium sulphite using an industrial process known as scrubbing, which involves directing the flue gases up a vertical absorber column whilst a liquid potassium sulphite solution flows down the column and absorbs sulphur dioxide. Other practical methods include the addition of additives (eg magnesium compounds) to the fuel, adsorption (ie the formation of a layer of SO_2 on the surface of the adsorbing material in contrast to the more complete permeation that occurs during absorption) and the oxidation or reduction of SO_2. One method uses extensive quantities of limestone which produces a gypsum by-product, for which there are fears that the market will soon become saturated. See also *integrated gasification and combined cycle technology; scrubber.*

fluidised bed

A combustion technology in which a sand bed is fluidised by vertical air jets, heated to temperatures high enough to support combustion, combustible material is added, and the process becomes exothermic. Fluidised bed combustion permits rapid and even heat transfer and can reduce generation of NOx. Materials such as lime can be added to the bed to absorb pollutants such as SOx. Fluidised beds are also used for chemical processes and filtration.

fluoridation

The addition of fluoride to public water supplies as a precaution against dental decay. Water containing the optimum amount of fluoride appears to increase resistance to tooth decay, while waters containing too much fluoride are capable of producing dental fluorosis which involves a mottling of the tooth enamel associated with brittleness and general deterioration. Fluoride occurs naturally in water and in rocks and soils. In a cool climate any potable water supply containing less than 0·8 gm per cubic metre of fluoride is considered deficient. The optimum usually recommended is 1·0 gm per cubic metre. The World Health Organisation regards fluoridization as a landmark in the history of public health. Investigation appears to affirm that wherever fluoridation has been introduced, dental decay in the teeth of children has been reduced by 65–75%.

fluorine

A non-metallic element belonging to the halogen group which is the most powerful oxidising agent known. It forms a gas of diatomic molecules which therefore reacts vigorously with most substances at room temperature, frequently with ignition. It forms fluorides with all elements except some of the noble gases, and most of these are stable compounds. It is reacted with crude oil derivatives to form a variety of fluorocarbons and chlorofluorocarbons, used for example as refrigerants, solvents, fire extinguishers and dielectric fluids. Some of these are volatile and cause removal of ozone from the upper atmosphere. Inorganic fluorides are regarded as essential for life in trace quantities but can give rise to toxic problems at higher concentrations, particularly to plants.

fly ash

Low density, finely divided material, carried over in the exhaust gases from the incineration of combustible matter. Pulverised fly ash (PFA) can be reused in building blocks.

fly-tipping

The illegal depositing or disposal of waste in an unauthorised manner and location. Special measures to counter fly-tipping were taken in the Control of Pollution (Amendment) Act 1989, under which all owners of vehicles transporting controlled waste *(qv)* in the course of a business or for profit are required to be registered (the certificate of registration to be carried in the cab at all times), and vehicles used for fly-tipping are liable to be seized and disposed of by the waste regulation authority (WRA) *(qv)*.

Food and Environment Protection Act 1985 (FEPA)

Legislation which conferred enabling powers for the Secretary of State for the Environment to make emergency orders dealing with contaminated food (Part I); replaced the controls which formerly operated under the Dumping at Sea Act 1974 concerning the depositing of substances and materials in the sea and made provision for controlling the depositing of substances and materials under the sea bed (Part II); also regulates standards in relation to pesticides (Part III).

food chain

A sequence in which nutritional energy is transferred from one organism to another; green plants (the primary producers) form the start of the sequence; each organism in the chain feeds on that below it in the sequence; plants are eaten by herbivores which are themselves eaten by carnivores which, in turn, may be consumed by carnivores higher up the chain. Since organisms often feed from several levels in the chain the concept has been developed into a more complex nutritional relationship called a food web.

food web

See *food chain.*

forestry

The management of forests. Defined for the purposes of the Rent (Agriculture) Act 1976 as including the use of land as nursery grounds for trees, and the use of land as woodlands where that use is ancillary to land use for other agricultural purposes. Statutory controls on felling of trees in forests are contained in the Forestry Act 1967.

The United Nations Conference on Environment and Development (UNCED) *(qv)* in 1992 adopted the Statement of Forest Principles for a global consensus on the management, conservation and sustainable development *(qv)* of all types of forest. The UK and other European countries are also committed to implementing the guidelines for the sustainable management of European forests, and for the conservation of their biodiversity, adopted at the Ministerial Conference on the Protection of European Forests in Helsinki in June 1993.

In 1995, forests were a key theme of the United Nations Commission on Sustainable Development (UNCSD) and there are other international agreements which bear more or less directly on their management, for example the Berne Convention *(qv)*, the Convention on International Trade in Endangered Species (CITES) *(qv)*, the General Agreement on Tariffs and Trade (GATT) *(qv)*, the International Plant Protection Convention and the International Tropical Timber Agreement.

The 1993 Helsinki Conference recognised the unique pressures on the forest resources of countries in transition to market economies, from emissions from old and highly polluting industrial plant, and from the demands on their natural resources created by the need to improve their economies and to generate foreign exchange. Although the EC does not have a common forestry policy, it does offer financial assistance for a number of forestry activities, such as preventing forest fires, monitoring tree condition, the marketing and processing of forest products and woodland improvement. Regulation 2080/92 provides financial assistance towards the creation of new woodlands and forests on agricultural land and requires Member States to prepare a national programme. The UK programme is based on existing national grant schemes (about £100 m) and EC aid to UK forestry (about £8 million a year). See *Sustainable Forestry: The UK Programme*, HMSO Cm 2429 (1994).

Forestry Commission

The Government Department with principal responsibility for forestry in England, Scotland and Wales (the Department of Agriculture in Northern Ireland). It has regulatory functions in relation to tree felling and the award of grants (carried out by its Forestry Authority arm) and has forest management functions in relation to state-owned forests (carried out by its Forestry Enterprise arm).

forward planning

As opposed to development control *(qv)*; the process in which national, regional and local planning policies are formulated by the Government and local authorities and recorded in development plans *(qv)* in order to guide the determination of applications for planning permission and other determinations under the Planning Acts.

fossil fuels

Materials which can be used as a source of energy (including coal, natural gas, petroleum, shale oil and bitumen) are found occurring naturally within the Earth's crust and were formed by the action of geologic processes on the remains of organic plant matter which grew hundreds of millions of years ago. All have a high carbon and hydrogen content as their chemical structure is based on carbon atoms. Fossil fuels provide the source for almost 90% of worldwide energy consumption.

free trade

Trade which is not regulated or restricted by way of customs duties, charges having equivalent effect, quantitative restrictions or any other means, and which is not hindered by differing technical standards, labelling requirements or any other trade barrier. See also *General Agreement on Tariffs and Trade (GATT).*

fresh waters

Waters containing only a small concentration of solubilised salts (usually inland surface waters such as waterways, rivers, lakes and lagoons); also used to describe waters which are not stagnant or which contain no impurities. Defined for the purposes of EC Directive 78/659 on fresh waters needing protection or improvement in order to support fish life as 'those running waters or standing fresh waters which support or which, if pollution were reduced or eliminated, would become capable of supporting fish belonging to indigenous species offering a natural diversity, or species the presence of which is judged desirable for water management purposes by the competent authorities of the Member States'.

Friends of the Earth

An international environmental campaigning organisation. [*Address: 26–28 Underwood Street, London, N1 7JQ; T: 0171 490 1555.*]

fugitive emissions

The escape of gaseous or particulate matter from process equipment or a confined/sealed system in an unintended and/or uncontrolled manner. There are three general categories: discontinuous emissions linked to specific operational processes (eg discharges occurring when hatchways are opened for inspection purposes); diffuse emissions arising from leakages (which may be continuous or intermittent); and emissions from open sources (such as stockpiles in windy conditions).

fumes

Gaseous emissions generally containing minute particles which may or may not form a cloud visible to the human eye; often formed by the condensation of vapours or as a product of a chemical reaction. The word is regularly used to refer to airborne emissions which are harmful or unpleasant. Part III of the Environmental Protection Act 1990 (EPA 1990) lists fumes as a category of emissions capable of creating a statutory nuisance and defines them as 'airborne solid matter smaller than dust' (s.79).

furans

Collective name for the group of chemicals called dibenzo furans (C_4H_40). A colourless, flammable liquid, relative density 0·94, melting point –85·6° C, boiling point 31·4° C. Readily soluble in alcohol, ether, acetone and benzene; insoluble in water. Decomposes in concentrated acids to form esters; stable in alkalis. Can be detected qualitatively by testing with spruce shavings (spruce wood dampened with hydrochloric acid will turn emerald green in the presence of furans). Used in the production of herbicides and pharmaceuticals. Has a chloroform-like odour; vapours are narcotic and can be absorbed through the skin. One of the volatile components in tobacco smoke. Lethal concentration in air for rats 30,400 ppm.

furnaces

Enclosed structures containing a heat source, typically used for the purpose of intense heating. Most are lined with refractory material, the heat source is typically provided by electrical elements or the burning of gas, coke or coal. Many different designs have been developed to perform a variety of functions including open hearth and basic oxygen furnaces (for steel production), blast furnaces (commonly used for iron smelting), electric arc furnaces, muffle furnaces, reverberatory furnaces, kilns and Mannheim furnaces.

Gaia

The Earth Goddess in Greek mythology, and the name given to the hypothesis of the planet Earth as a living organism in which homeostasis is maintained by active feedback processes operated automatically and unconsciously by the biota. Further reference: James Lovelock, *Gaia: A New Look at Life on Earth* (1982) and *The Ages of Gaia* (1988).

gas chromatography (GC)

The process in which the components of a mixture are separated from one another by volatilising the sample mixture into a carrier gas stream which passes through and over a packing in a column. If the packing is a solid the process is called gas-solid chromatography or vapour phase chromatography (VPC); if the surface of the packing is coated with a relatively non-volatile liquid the process is called gas-liquid chromatography (GLC). A chronographic output is obtained by siting a suitable detector at the effluent end of the column. One method uses a mass spectrometer as the detection system (GC/MS). Gas chromatography is a particularly useful technique for measuring trace substances in water, soil and air samples taken from the environment.

General Agreement on Tariffs and Trade (GATT)
A treaty (concluded in 1948) which seeks to establish a uniform code of conduct for the carrying on of international trade through a variety of means, including the reduction of tariff barriers and the outlawing of protectionist measures by which nations support their domestic industries. GATT aims to promote the trading performance of developing countries, which are frequently subject to less stringent controls than developed areas. Disputes are heard by specialist panels, one of which recently determined that the US was not entitled to close its market to tuna caught in drift nets. At present, GATT governs around 90% of the world's trade; the terms on which trade is regulated are negotiated in long-running conferences known as rounds. The most recent, the Uruguay Round, was concluded with some difficulty in December 1993, and came into effect on January 1, 1995. It will extend GATT into new fields such as textiles, services and intellectual property rights. The Uruguay Round envisaged the setting up of the World Trade Organisation as the administration for GATT.

General Development Order (GDO)
The Town and Country Planning (General Permitted Development) Order 1995 (SI 1995 No 418) is a statutory instrument *(qv)* made under the Town and Country Planning Act 1990 which grants 'permitted development rights', amounting to an automatic national grant of planning permission *(qv)* for specified classes of development *(qv)*. The Order contains 33 categories of permitted development, notably those for 'householder' development (Part 1); agricultural development (Part 6); industrial development (Part 8) and various categories of development by public bodies and utilities. It is possible for the local planning authority *(qv)* to suspend the operation of the GDO in defined areas by making an Article 4 direction *(qv)* with the result that planning permission must be sought for the development. The 1995 Order supersedes the Town and Country Planning General

Development Order 1988, and the procedural provisions of the 1988 Order were transferred into a separate instrument, the Town and Country Planning (General Development Procedure) Order 1995 (SI 1995 No 419). The equivalent measures for Scotland are the Town and Country Planning (General Permitted Development) (Scotland) Order 1992 (SI 1992 No 223) and the Town and Country Planning (General Development Procedure) (Scotland) Order 1992 (SI 1992 No 224).

genetically modified organism (GMO)

Defined in the Environmental Protection Act 1990 (EPA 1990) s.106 as any acellular, unicellular or multicellular entity (in any form) other than human or human embryos, including substances consisting of or including either:

1 tissues or cells or sub-cellular entities capable of replication or of transferring genetic material; or

2 genes or other genetic material in any form which are capable of so doing, where any of the genes or genetic material in the organism have been modified by an artificial technique or where they are inherited or otherwise derived (through any number of replications) from genes or genetic material which was so modified.

In view of the potential impact of GMOs on the environment and on human beings, the import, acquisition, keeping, release or marketing of GMOs is regulated by Part VI of the Act and by regulations made thereunder, particularly the Genetically Modified Organisms (Contained Use) Regulations 1992 (SI 1992 No 3217) and the Genetically Modified Organisms (Deliberate Release) Regulations 1992 (SI 1992 No 3280).

global environmental facility (GEF)

A fund established in 1991 to promote environmental and conservation projects in developing countries, administered by the World Bank, the United Nations Development Programme and the United Nations Environment Programme *(qv)*. Established for a pilot period of three years with a commitment from donors of about $1 bn, the GEF was restructured in April 1994 for a further two years with a further sum of $2 bn for the period 1994–97.

global warming

The phenomenon that scientists predict will occur as a result of uncontrolled exaggeration of the greenhouse effect *(qv)*. According to the Intergovernmental Panel on Climate Change in its *First*

Assessment Report 1990 (Cambridge University Press, 1990):

1 the concentration of greenhouse gases in the atmosphere has increased substantially as a result of human activity;

2 this is expected to enhance the natural greenhouse effect, which keeps the earth warmer than it otherwise would be;

3 average global temperatures have increased by 0·3°C to 0·6°C during the last century;

4 without action to restrain emissions, an increase in global average temperatures of around 0·3 per decade (with an uncertainty range of 0·2°C to 0·5°C) is likely in the future.

This could imply sea level rise of around 6 cm per decade (with an uncertainty range of 3 to 10cm). A second report by the panel in 1992, *Climate Change in 1992* (Cambridge University Press, 1992) confirmed the earlier findings, though still stressing the levels of uncertainty, and that it will be some years before the warming effect moves beyond the bounds of natural climate variability.

'Going for Green'
The adopted name for the Citizens' Environment Initiative *(qv)*.

golf courses
This is not strictly an environment term but of interest. Environmentalists claim that golf courses can have a significantly deleterious effect on the environment unless properly planned and maintained. Pesticides used to keep fairways fair and greens green often flow into nearby streams killing fish and poisoning the nearby tap water. Clumsy excavation by golf course contractors causes landslides such as the one that killed 54 people in South Korea in 1992. Developers in poor countries steal water from irrigation projects to use on the courses, leaving local paddy fields parched.

The anti-golf lobby would contain fewer environmentalists if all golf course architects were as sensitive as those in Scotland, the home of golf. There, the aim has always been to design courses that use the existing contours of the land, rather than moving thousands of tonnes of earth as occurs in the US or installing escalators between sunken greens and raised tees, as in Japan. Most of the top 30 courses in Britain are either Scottish or built along Scottish lines and fully half of them have parts designated sites of special scientific interest (SSSIs) *(qv)* by the Nature Conservancy Council.

Golf clubs in England and Wales possessing such courses are able to argue convincingly that they are more friendly to the environment than many other developments. A fern or heather flanked fairway certainly nurtures more birds and insects than the car park of an urban shopping development. A golf course is arguably just as eco-friendly as an arable field that is treated with fungicides, herbicides and pesticides *(qv)*.

Green Alliance

An independent non-profit making organisation working on environmental policy, which aims to raise the prominence of the environment on the agendas of all key policy-making institutions in the UK, and to help improve environmental performance across the board. [*Address: 49 Wellington Street, London, WC2E 7BN; T: 0171 836 0341; F: 0171 240 9205.*]

green belt

An area defined in a development plan *(qv)* for protection from development, with the overall policy objective of preventing urban sprawl. There is a strong presumption against development being permitted in a green belt, and the construction of any building is regarded as inappropriate unless it is for agriculture or forestry, essential for outdoor sport and recreation, involves the limited extension of existing dwellings, limited infilling in existing villages or infilling or redevelopment on sites already developed and identified in adopted local plans. Approved green belts now cover over 1·5m ha in England, about 12% of its areas. Further reference: PPG2 (revised), *Green Belts* (1995).

green dot

See *Duales System Deutschland (DSD)*.

greenhouse effect

A natural phenomenon which helps to maintain the Earth's surface at a temperature suitable for sustaining life by preventing accumulated heat from escaping into space (thus the analogy with the greenhouse). The Earth's atmosphere acts as a filter which controls the exchange of energy between the sun, the Earth and space. Certain gases in the atmosphere allow short-wave solar radiation in the form of visible or ultraviolet light to pass through this filter to the Earth's surface, while absorbing the long-wave infra-red radiation reflected back from the surface in the form of heat. These heat-retaining gases are the so-called greenhouse gases *(qv)*, and it is the proliferation of such thermal catalysts in the Earth's atmosphere as a result of the development of human activity which may have caused an uncontrolled and apparently harmful exaggeration of the greenhouse effect leading to global warming *(qv)*. The greenhouse effect has become a global concern. In 1992, the Rio Summit *(qv)* adopted the Climate Change Convention *(qv)*, with the objective of stabilising greenhouse gas concentrations and preventing dangerous interference with the climatic system.

greenhouse gases

Those gaseous constituents of the atmosphere, both natural and anthropogenic, that absorb and re-emit infrared radiation. Such gases have the property of being nearly transparent to incoming radiation from the sun, while retaining some of the energy re-emitted by the earth as long wavelength infrared radiation.

The outcome of this mechanism is that part of the radiant energy coming from the sun is trapped in the lower atmosphere, the so-called 'greenhouse effect' *(qv)*. The principal (by volume) anthropogenic greenhouse gas is carbon dioxide (CO_2), followed by chlorofluorocarbons (CFCs) *(qv)*, methane (CH_4), and nitrous oxide (NO_2). Atmospheric water vapour is also a greenhouse gas.

greening the GATT

The General Agreement on Tariffs and Trade (GATT) *(qv)* broadly agrees trade barriers and quotas between member countries at irregular intervals during a process called the GATT rounds. Environmentalists consider the results of the most recent Uruguay Round, finished under some frenetic bargaining pressure, ignored necessary green factors since the round barely touched on the environment. So there has been talk of a green round to redress the balance on the basis that trade, in this context, touches the environment in several ways, for example:

- trade brings economic growth, which can damage environments, especially where prices do not reflect environmental costs
- national environmental policies sometimes conflict with commitments made under GATT
- countries might want to use trade sanctions as a threat to wrest better environmental standards from other countries
- trade could enable polluters to use countries of lax environmental rules as export platforms to markets where standards are higher. This would not only penalise conscientious firms but make it harder for environmentalists to raise standards.

In recommending policies to respond to these concerns, environmentalists and free traders are not necessarily at one. Environmentalists consider that the undemocratic trading system must be overhauled whatever the cost and more comprehensive attempts be made to steer between the Scylla of blind environmentalism and the Charybdis of narrowly focused trade liberalisation.

Under existing GATT rules, a country cannot use trade sanctions to influence manufacturing processes, however polluting they may prove to be. As long as a product meets an importer's standard, it is irrelevant that it was made using banned chemicals. That is the dilemma, hopefully to be resolved over the next 20 years by greening the GATT, albeit to a limited extent, during the next GATT rounds.

Greenpeace

An international environmental organisation which conducts high-profile campaigns on selected issues and also commissions research into environmental problems. Brought proceedings for judicial review *(qv)* against Her Majesty's Inspectorate of Pollution (HMIP) *(qv)* in relation to the commissioning of the Thorp nuclear reprocessing plant at Sellafield, Cumbria, which were eventually unsuccessful, but none the less secured an unsolicited judicial endorsement that: 'British Nuclear Fuels Ltd rightly acknowledges the national and international standing of Greenpeace and its integrity. So must I. I have not the slightest reservation that Greenpeace is an entirely responsible and respected body with a genuine concern for the environment' (Otton J, in *R v HMIP (No 2), ex parte Greenpeace Ltd* [1994] 4 All ER 329, at 350). [*Address: Canonbury Villas, London, N1 2 PN; T: 0171 354 5100.*]

grey list

A colloquial term applied, in the context of a number of national and international law instruments, to the list of substances whose discharges that instrument does not seek to prohibit entirely, but which should be reduced to the extent possible. A familiar example is List II of the Annex to EC Council Directive 76/464 on pollution caused by certain dangerous substances discharged into the aquatic environment of the Communities.

ground-level concentration

The amount of pollutant found in a band of air from ground level to about 2 metres above it, ie a band of air to which a human being may be exposed. The concentration of pollutant should not be measured at ground level because some of it may have been absorbed by the ground. Because of local effects of changing wind speed and direction it is unwise to rely on one measurement and it is better to obtain an average concentration over a given length of time.

ground-penetrating radar (GPR)

A technique for acquiring subsurface information by the downward emission of pulses of high frequency electromagnetic waves, which are reflected back to receiving antennae. Soil and rock layers with sufficiently different electrical properties are shown in a continuous cross-sectional profile. GPR is useful in hazardous waste site assessments and prospecting, allowing the evaluation of natural soil and geologic conditions; the location and delineation of buried waste materials and contaminant plume areas; and the location and mapping of buried utilities, both metallic and non-metallic.

groundwater

That part of the natural water cycle which is present within underground strata, or aquifers *(qv)*: 'all water which is below the surface of the ground in the saturation zone and in direct contact with the ground or subsoil' (EC Directive 80/68 on the protection of groundwater against pollution caused by certain dangerous substances). The importance of groundwater resources as a major source of high-quality water for public use and consumption has long been recognised in the UK.

Groundwater is vulnerable to pollution from diffuse sources, particularly the migration of contaminated surface waters or through leachate *(qv)* from industrial undertakings or landfill sites. Once contaminated, underground waters cannot easily be made wholesome again. Because of the slow movement of groundwater through the strata, pollution may take many years to become manifest.

The Directive defined the practices acceptable for the protection of groundwater, such as the prohibition of any activity if it would lead to the discharge of certain substances in significant amounts from List 1 or the authorisations necessary for the control of List 2 substances. These practices are likely to be updated in an amending draft Directive on groundwater protection expected to be published in 1995. In the meantime, the National Rivers Authority (NRA) has established a set of policies for the protection of groundwater using its own statutory powers and also relying on planning control *(qv)*. The strategy includes the definition and protection of source protection zones *(qv)*. Further reference: NRA, *Policy and Practice for the Protection of Groundwater* (1992).

Groundwork Trust

A UK organisation that assists people from all sectors of local communities to improve their local environment through partnership. There are now 34 trusts, covering over 80 local authority areas.

guide standard

A standard specified in EC legislation as a recommended standard, more stringent than the minimum mandatory standard which may be set in the same legislation. See also *imperative standard.*

habitat

The normal locality in which a species or community of plant or animal naturally lives and grows.

Habitats Directive

Council Directive 92/43 on the conservation of natural habitats and of wild fauna and flora which is intended to assist in the conservation of biological diversity within the Member States of the European Communities (EC). The Directive seeks to ensure that a wide variety of habitats within the Member States remain at, or where necessary are restored to, a level at which their conservation status can be described as favourable. In particular, the Directive envisages the identification of a network of special areas of conservation (SACs) *(qv)*, which will make up Natura 2000 *(qv)*. Member States are obliged to submit a list of eligible sites to the EC, which will then establish a schedule of sites of importance on the EC level which will be reviewed and finalised by a panel of experts from the Member States. Member States are placed under particular obligations in respect of SACs, notably to take steps to avoid their deterioration in quality.

half life

The time taken for one half of the atoms of a radioactive isotope to disintegrate. The half life for iodine 131 is 8 days, for strontium 90 about 28 years, for caesium 137 about 30 years, and for radium about 1,580 years.

halogens

The collective term given to the five elements of Group VIIA of the periodic table: fluorine, chlorine, bromine, iodine and astatine. Halogens are electronegative and therefore are strong oxidising agents. They are classified as non-metals. Their inorganic compounds readily form negatively charged ions in solution. They combine with organic compounds to form products such as fluorocarbons, chlorinated hydrocarbons and various plastics.

halons

Compounds of methane, bromine, fluorine and sometimes chlorine. They are chemically stable, enabling them to carry chlorine and bromine into the stratosphere, with ozone-depleting potential which exceeds that of chlorofluorocarbons (CFCs) *(qv)*. Halons are typically used as fire extinguishants. The aim of the Montreal Protocol *(qv)*, to which the UK is a party, is to set targets for the phasing out of halons and other ozone-depleting substances.

hazardous substance

A specific term in UK environmental law, meaning a substance which has been designated as hazardous by the Secretary of State for the Environment and appears in Schedule 1 to the Planning (Hazardous Substance) Regulations 1992 (SI 1992 No 656). A different list (compiled for notification purposes) appears at Schedule 1 to the Notification of Installations Handling Hazardous Substances Regulations 1982 *(qv)*. All substances can be potentially hazardous; it is the particular circumstances which render a substance or product hazardous. A hazard is a potential source of possible harm and the risk is the likelihood of that harm being realised. The contents of a tanker labelled 'non-hazardous substance' could in a commonplace accident cause harm to a child, or to the ecosystem of a road-side stream.

Hazardous Substances Authority

The local authority with responsibility for administering the controls over hazardous substances conferred by the Planning (Hazardous Substances) Act 1990. Usually, the district or unitary council for the land in respect of which a hazardous substance consent *(qv)* is required, but the county council where the land is situated in a non-metropolitan county and is used for mineral extraction or is a waste disposal site (in England). Where the land is in a National Park, the hazardous substance authority is the joint (or special) planning board, but if there is no such board, the county council.

hazardous substances consent

A specific consent which is required in order that a hazardous substance *(qv)* may lawfully be present on any land in excess of certain threshold quantities. Application for consent is made to the appropriate hazardous substances authority *(qv)* under the Planning

(Hazardous Substances) Act 1990. Hazardous substances in transit do not require a consent, unless they are unloaded, but they do require proper packaging and labelling in a suitable vehicle conducted by a trained driver/operator.

hazardous (toxic) waste

Not strictly a technical term in UK law. Attempts to define the term by reference to substances or properties have bedevilled the efforts of a number of international organisations. For practical purposes, probably the definitive list is to be found in the Annexes to EC Council Directive 91/689 on hazardous waste as elaborated by Council Decision 94/906 (December 22, 1994). This Directive (as amended by 94/31) attempts to establish controls appropriate for the most potentially harmful wastes as elaborated by Council Decision 94/906 (December 22, 1994). This is to be implemented in the UK by a revision of the Special Waste Regulations 1980. The Directive includes a new definition of hazardous waste by reference to a waste list which has proved very difficult to compile. Issue of a final version of this list was delayed at least twice but had to be adopted by December 27, 1994 in readiness for the Directive's revised implementation date in the UK, and the 14 other countries, of June 27, 1995.

A US definition is provided by the Resource Conservation and Recovery Act 1976 (RCRA) *(qv)* which defined hazardous waste as a solid waste or combination of solid wastes, which because of its quantity, concentration or physical, chemical or infectious characteristics, may:

1 cause or significantly contribute to an increase in mortality; or

2 increase in serious irreversible, or incapacitating reversible illness; or

3 pose a substantial present or potential hazard to human health or the environment when improperly treated, stored, transported or disposed of, or otherwise managed.

health and safety at work

The body of laws and practice which regulate the duties of employers to ensure the health and safety of employees (and others) in the workplace. Much of the statute law on the subject was codified in the Health and Safety at Work Etc. Act 1974 and in the many regulations made under it. This statutory regime is backed up by criminal sanctions and, in relation to breaches of regulations, provides the basis for civil claims for compensation. An important role in the description of the standards to be observed is played by Approved Codes of Practice (ACOPS) *(qv)*, typically drafted by the Health and Safety Executive *(qv)*. Many aspects of health and safety at work abut and are complementary to environmental law. Some environmental legislation provides expressly that conditions on environmental consents shall not be imposed solely for the purpose of protecting the health and safety of persons at work. Obvious examples are labelling, packaging, storage and use of chemicals and other dangerous substances, indoor air pollution *(qv)* and certain powers of the Health and Safety Executive to take action in respect of activities which pose public health risks off-site.

Health and Safety Commission (HSC)

A supervisory body established under the Health and Safety at Work Etc. Act 1974 to oversee the activities of the Health and Safety Executive.

Health and Safety Executive (HSE)

A regulatory agency established under the Health and Safety at Work Etc. Act 1974 to enforce and implement the law relating to the avoidance of threats to the health and safety of employees (and others), whether in the workplace or arising from it. The HSE has a wide remit, not all of which is directly related to environmental law, but areas which do significantly elide with it include the law relating to exposure to asbestos, the control of dangerous substances, activities which may give rise to major hazards (within the meaning of the Seveso *(qv)* Directive) and the Control of Substances Hazardous to Health Regulation 1994 (COSHH) *(qv)*. The HSE also has some functions in relation to nuclear installations through the medium of its Nuclear Installation Inspectorate. The HSE has power to serve improvement notices in respect of hazards in the workplace requiring the addressee to take steps to remove the hazards, whether by installing new equipment or by adopting enhanced safety procedures. The HSE may also serve a prohibition notice directing that an activity should cease (in the case of a prohibition notice, the HSE may take action only when it anticipates the possibility of injury in the workplace itself or similar danger to the public beyond the factory fence).

heavy metals

A chemically imprecise term which is nevertheless very widely used. It is usually applied to non-ferrous metals, mainly but not exclusively transition elements, with partly filled d or f electron shells, many of which may be harmful to human health or the environment. Examples include arsenic, antimony, cadmium, chromium, copper, lead, mercury, nickel, selenium and zinc.

hedgerow

A form of managed natural fencing comprising shrubs and saplings, interwoven so as to prevent the escape of livestock. The widespread destruction of hedgerows that has accompanied contemporary mechanisation of agriculture has transformed the English countryside and caused wide concern. There is no general legal protection for hedgerows, since their destruction does not require planning permission *(qv)*; nor can hedgerows be the subject of a tree preservation order since their components are not normally of sufficient size to constitute trees. However, the Environment Bill 1995 proposes a new regime for the protection of 'important hedgerows', to be prescribed wholly by regulations.

Helsinki Convention

See *Convention on the Transboundary Effects of Industrial Accidents.*

Her Majesty's Industrial Pollution Inspectorate (HMIPI)

The principal agency for the implementation of integrated pollution control *(qv)* in Scotland, with functions broadly equivalent to those of Her Majesty's Inspectorate of Pollution *(qv)* in England and Wales. Appointments to the Inspectorate are the responsibility of the Secretary of State for Scotland. Under the Environment Bill 1995 *(qv)* the Inspectorate will be absorbed in the new Scottish Environmental Protection Agency.

Her Majesty's Inspectorate of Pollution (HMIP)

A central Government environmental agency formed from the amalgamation in 1989 of a number of specialised inspectorates, including the Industrial Air Pollution Inspectorate (formerly the Alkali Inspectorate). HMIP is responsible for granting authorisations under the system of integrated pollution control *(qv)* established under the Environmental Protection Act 1990 Part I; for regulating the use of sites on which radioactive material is used, stored or disposed of under the terms of the Radioactive Substances Act 1993; for regulating the discharge into sewers of red list *(qv)* substances under the Water Industry Act 1991 and for carrying out research into pollution control and advising the Government, other regulatory agencies and industry on pollution control issues. Although a central Government body, it is organised regionally. Its remit does not extend to Scotland or Northern Ireland. Under the Environment Bill 1995 *(qv)*, HMIP will be subsumed into the new Environment Agency.

holistic (approach)

An approach based on an appreciation that each individual and, superficially separate, element in a system must be considered as part of the whole. Originally used by critics of post-WW2 materialistic, consumption-based societies.

House of Commons Select Committee on the Environment

A committee of the House of Commons that was established in 1982 to monitor the performance of the Department of the Environment. It has power to send for persons, papers and records and to question witnesses, and to report to the House. The Government replies to each report with a statement of reasons for accepting or rejecting its recommendations. The House of Commons Select Committee on the Environment has had a large impact on the direction of environmental policy, especially under its first chairman, Sir Hugh Rossi MP. The Committee pressed in several reports for the establishment of a national environment agency, which was initially resisted by the Government but now is the central proposal of the Environment Bill 1995 *(qv)*. The Committee chooses its own topics for deliberation, and invites the submission of written and oral evidence. The Minister responsible for the particular area of policy is normally questioned orally by the Committee in public session. Its reports have influenced Government thinking.

The principal reports of the Committee on environmental issues include:

Session 1983–84

Acid Rain (HC 446) (Government reply published as Cmnd 9397)

Session 1984–85

Operation and Effectiveness of Part II of the Wildlife and Countryside Act 1981 (HC 6) (Government reply published as Cmnd 9522)

Session 1985–86

Follow-up to the Environment Committee Report on Acid Rain (HC 51)

Radioactive Waste (HC 191) (Government reply published as Cmnd 9852)

Session 1986–87

3rd Report: Pollution of Rivers and Estuaries (HC 183) (Government reply published as HC 543 (1987–88))

Session 1987–88
1st Report: Air Pollution (HC 270) (Government reply published as Cm 552)

Session 1988–89
1st Report: Registration of Carriers of Controlled Wastes: Recommendation regarding the Control of Pollution (Amendment) Bill (HC 222)
2nd Report: Toxic Waste (HC 22) (Government reply published as Cm 679)
7th Report: The Proposed European Environment Agency (HC 612)

Session 1989–90
1st Report: Contaminated Land (HC 170) (Government reply published as Cm 1161)
2nd Report: European Community Environmental Policy (HC 372)

Session 1990–91
7th Report: The EC Draft Directive on the Landfill of Waste (HC 263) (Government reply published as Cm 1821)
8th Report: Eco-labelling (HC 474) (Government reply published as Cm 1720).

Session 1991–92
1st Report: The Government's Proposals for an Environment Agency (HC 55)
2nd Special Report: Review of the Committee's Work 1983–1992 (HC 340)

Session 1992–93
1st Special Report: The Government's Proposals for an Environment Agency (HC 256)

Session 1993–94
2nd Report: Recycling (HC 73)

House of Lords Committee on Sustainable Development
A committee established by the House of Lords in 1994 to carry out a special investigation into the Government's progress towards sustainability, and due to report in May 1995.

House of Lords Select Committee on European Communities
A committee which is appointed at the beginning of each session of the House of Lords to consider proposals coming from the European Communities (EC), to obtain all necessary information about them and

to make reports on those which in its opinion raise important questions of policy or principle and on other matters to which the Committee considers that special attention of the House of Lords should be drawn. The Committee has powers to appoint sub-committees, and one of these has been its Environment Sub-committee. The Committee also has power to exchange papers with the equivalent committee in the House of Commons, and to sit concurrently with the House of Commons committee or one of its sub-committees. The Environment Sub-committee has undertaken several major investigations into aspects of European environmental policy and their implications for the United Kingdom, including the following:

Session 1986–87
4th Environmental Action Programme (HL Paper 43)

Session 1988–89
Municipal Waste Water Treatment (HL Paper 73)

Session 1989–90
Freedom of Access to Information on the Environment (HL Paper 2)
Paying for Pollution (HL Paper 84)

Session 1990–91
Municipal Waste Water Treatment (HL Paper 50)
Energy and the Environment (HL Paper 62)

Session 1991–92
Carbon Energy Tax (HL Paper 52)

Session 1992–93
Control of National Treasures (HL Paper 17)
5th Environmental Action Programme: Integration of Community Policies (HL Paper 27)
Environmental Aspects of the Common Agricultural Policy (HL Paper 45)
Implementation and Enforcement of Environmental Legislation (HL Paper 53)
Industry and the Environment (HL Paper 73)
Packaging and Packaging Waste (HL Paper 118)

Session 1993–94
Remedying Environmental Damage (HL Paper 10)
A Community Eco-audit Scheme (HL Paper 42)
Common Transport Policy – Sustainable Mobility (HL Paper 50)
Protection of Wild Birds (HL Paper 70)

Session 1994–95
Bathing Water (HL Paper 6)

hybrid Bill
A parliamentary Bill introduced as a public Bill, normally by the Government, which contains provisions affecting certain individuals or corporations in some specific way, usually in terms of their ownership of land or other property.

If declared hybrid, a Bill must follow a special procedure in both Houses of Parliament, including the holding of quasi-judicial hearings of objections lodged by objectors with *locus standi (qv)*, at which environmental and other issues will be evaluated. A procedure sometimes used to authorise major construction projects where the alternative would be a public local inquiry or (formerly) a private Bill. Examples include the Channel Tunnel Act 1987 and the Channel Tunnel Rail Link Bill introduced to Parliament in the 1994–95 session. See also *private Bill procedure.*

hydraulic head
The difference in level necessary to produce flow in a liquid.

hydrobromofluorocarbons (HBFCs)
Substances containing carbon, hydrogen, fluorine and bromine, each with ozone-depleting properties. Under Article 2G of the Montreal Protocol *(qv)* on substances that deplete the ozone layer both the production and the consumption of HBFCs are required to cease altogether from January 1, 1996.

hydrocarbons
Organic compounds consisting of carbon and hydrogen, found commonly in fossil fuels and the products of incomplete combustion of these, eg vehicle exhaust fumes.

hydrochlorofluorocarbons (HCFCs)
Compounds of hydrogenchlorine, fluorine and carbon which are chemically similar to chlorofluorocarbons (CFCs) *(qv)* and are often used as substitutes in refrigeration, foam blowing and aerosols. Less active as ozone depletors than CFCs, but still controlled as transitional substances under the Montreal Protocol *(qv)*; and are agents of global warming *(qv)*.

hydrofluorocarbons
Halogenated carbons, similar to HCFCs, but not containing chlorine, and therefore not damaging to the ozone layer *(qv)*.

hydrogen sulphide

A colourless gas (formula H_2S), with a characteristic foul odour of rotten eggs and density greater than air, arising from the decomposition of organic material. Other sources include sulphur recovery plants and various chemical industries, oil refineries and some metallurgical processes. It is irritating to the eyes and respiratory tract, and can cause death through paralysis of the respiratory centres of the brain. Although low concentrations are easily recognisable by smell, olfactory fatigue occurs quickly at high concentrations making this an unsafe method of detection. Hydrogen sulphide is corrosive to many metals and discolours lead paints even at low atmospheric concentrations. In ordinary combustion processes hydrogen sulphide is readily burned to sulphur dioxide.

hydrograph

A graph showing the level of water, such as in a watercourse or well; or the rate of flow of water through time.

hydrology

The study of water levels and balances. The hydrological cycle is the continual exchange of water between the Earth and the atmosphere.

hydrolysis

A chemical or biochemical reaction in which a compound is split into two or more smaller molecules, with the addition of the elements of water. An example of hydrolysis of an inorganic compound is the reaction of ferric chloride, $FeCl_3$, with water, producing ferric hydroxide, $Fe(OH)_3$, and hydrochloric acid, HCl. An example of hydrolysis of an organic compound is the biologically mediated conversion of lipids into glycerol and long-chain fatty acids.

hydroponics

The technique of growing crops in an aqueous nutrient solution without soil.

imperative standard

A standard prescribed by the European Communities (EC) as a mandatory minimum standard (eg as to emission limits) to be set by Member States, as compared with a guide standard *(qv)*.

impoundment

The obstruction or containment of waters, typically by the construction of a dam or a weir. Impounding works in relation to controlled waters *(qv)* require a licence from the National Rivers Authority, under the Water Resources Act 1991, Part II, in accordance with a licensing scheme that follows closely that for abstraction *(qv)* of water.

incineration

A method of treatment and disposal of waste *(qv)* by combustion, which can also be used to recover energy from certain types of waste. The incineration process generally involves two stages. The waste is first burned at a sufficiently high temperature to turn some of the substances present into gases and release others as an aerosol of fine particles. In the second stage the mixture of gases and particles is burned at a higher temperature. Non-combustible material, the burnt-out remains of combustible material and fine particles carried out of the combustion chamber are collected as solid residues from the process. The Royal Commission on Environmental Pollution *(qv)* recommended in its 17th Report (1993) on *Incineration of Waste* that incineration with energy recovery should be expanded in the United Kingdom, a recommendation which was broadly accepted by the Government in its response to the report.

In England and Wales incineration is controlled by public authorities in the following four ways:

1 by the local planning authority *(qv)* as a development of land;

2 by the waste regulation authority *(qv)* as a form of waste management;

3 by Her Majesty's Inspectorate of Pollution (HMIP) *(qv)* or, for smaller plants, by the district council and other authorities, as a source of pollution; and

4 by the Health and Safety Executive *(qv)* where there is a potential hazard to workers on the site and persons outside it. In Scotland the legislation is broadly similar and in Northern Ireland the same principles are applied although the legislation is different.

indoor air pollution

A colloquial term which has gained currency in the wake of

heightened awareness of the need to have regard to air quality inside buildings and other artificial structures, especially in respect of legionella, radon *(qv)* and mineral fibres such as asbestos *(qv)*. See also the House of Commons Select Committee 6th Report (Session 1990–91), on *Indoor Pollution* (HC61; 1991).

Industry Council for Packaging and Environment (INCPEN)

An influential council established in 1974 to carry out research on the environmental and social effects of packaging. It brings together all sectors of the industry involved with packaging, including raw material suppliers, packaging manufacturers and manufacturers and retailers of packaged goods. It currently has 70 member companies and conducts research on packaging in areas such as raw material and energy use, lifecycle analysis, recovery of wastes, wastes management, litter, biodegradable materials, refillable container systems, economic instruments, and analysis of the effect of legislative measures.

inert

Inactive or unreactive. Used as a description of substances which usually have little or no chemical affinity or activity. For example, inert gases are helium, neon, argon, krypton, xenon and radon – 222. Of these, helium, neon and argon have a valence of 0 and do not enter into any chemical combination. The others are capable only of limited compound formation. The term 'inert atmosphere' refers to carbon dioxide and nitrogen, which are unreactive under normal conditions. Examples of inert solids are uncontaminated clays, asbestos, talc and sand. Not a descriptive term which should be generally applicable to waste materials.

The 1994 Waste Management Paper *(qv)* No 4 Licensing of Waste Management Facilities states that many sites were licensed under the Control of Pollution Act 1974 to accept inert wastes: experience has shown that inert was a misnomer. A very high proportion of these sites contained slowly degrading materials, such as wood from demolition wastes, that subsequently gave rise to the production of landfill gas and leachate.

infiltration

The process by which a fluid penetrates a solid, and commonly applied to the unintended ingress of groundwater into a drainage system.

information, environmental

Traditionally there has in the United Kingdom been only a limited right of public access to environmental information held by

Government or other public authorities, but this custom of official secrecy has come to be significantly undermined in recent years. In particular, the Environmental Information Regulations 1992 (SI 1992 No 3240), which implement EC Directive 90/313 on freedom of access to information on the environment, establish a general public right of access to information on the environment held by 'relevant persons'. 'Relevant persons' includes Ministers of the Crown, Government departments, local authorities and certain categories of bodies with public administration functions or public responsibilities in relation to the environment. Access to environmental information has also been improved through the publication of the Chemical Release Inventory (CRI) *(qv)* and the now largely standard statutory requirement to place on public registers information relating to authorisations, consents and licences (in planning control and under all other systems of environmental regulation). However, exceptions remain, particularly relating to national security and to commercial confidentiality (see eg Environmental Protection Act 1990, ss. 21 and 22).

innocent landowner defence

A defence which a landowner may mount against a claim for contribution to clean-up costs under US Superfund *(qv)* legislation. The legislation originally imposed strict liability *(qv)* on all owners of contaminated sites, but that was mitigated in the 1986 amendments which introduced this defence. None the less, the landowners 'must have undertaken, at the time of acquisition, all appropriate inquiry into the previous ownership and uses of the property consistent with good commercial or customary practice in an effort to minimise liability'. Innocent owners who learn of releases at the site, and then transfer the ownership to subsequent purchasers without disclosing the information, are fully liable, so that unless a clean-up is undertaken at someone else's expense, the 'innocent landowner' who subsequently acquires knowledge may still end up bearing the cost, either through undertaking clean-up himself or through having to sell the property on at an appropriately discounted price. Under proposals contained in the Environment Bill 1995, liability for clean-up of contaminated sites in the United Kingdom will first be channelled to the person who caused or knowingly permitted the substances to be there in the first place; but if that person cannot be found, it passes to the landowner (or occupier) for the time being (although in recovering any expenses the enforcing authority will be required to have regard to any hardship it may cause to that person).

insecticide

A biocide *(qv)* having the effect of or employed in the killing or control of insects.

Institute for Environmental Assessment (IEA)

A membership and business organisation established in 1990 to promote best practice standards in environmental assessment and auditing, offering quality reviews of environmental statements. It also advises and informs the industry on environmental consultants' skills and experience. The Institute organises annual conferences, meetings and training seminars and is the publisher of the quarterly *Environmental Assessment Review*. [*Address: Unit 6, The Old Malthouse, Spring Gardens, London Road, Grantham, Lincs; T: 01507 533 444.*]

Institute of Environmental Health Officers (IEHO)

Founded in 1983 and in 1984 was incorporated by Royal Charter. There are over 7,000 members, most of whom work for local authorities in England, Wales and Northern Ireland. The Institute regulates the training of environmental health officers *(qv)*, the qualification for which is the Diploma in Environmental Health or a degree which carries exemption from the Diploma examination. [*Address: Institute of Environmental Health Officers, Chadwick House, 41 Rushworth Street, London, SE1 0QT.*]

Institute of Environmental Management (IEM)

A professional association, formerly the Institute of Environmental Managers, founded in 1992 and open to individuals involved in environmental management in industry, commerce and local government. The Institute runs a best practice programme, dealing with such matters as the establishment of workable objectives and targets, the determination of significant environmental effects and motivation for environmental change in the workplace.

There are varying grades of Membership, culminating in Full Membership when the candidate has fulfilled the necessary accreditation standards. Fellowship of the Institute is awarded on a merit basis to individuals who have demonstrated substantial achievement in environmental management. All Members of the Institute are required to subscribe to the Environmental Managers' Charter, which serves as a code of professional conduct. [*Address: 58/59 Timber Bush, Edinburgh, EH6 6QH; T: 0131 555 5334; F: 0131 555 5217.*]

Institute of Wastes Management (IWM)

A scientific and technical organisation and a professional body, created with the objective to promote all scientific, technical and practical aspects of waste management. It provides advice, information, guidance and training and liaises with the Government. It also organises annually a national conference and exhibition and is generally growing in strength and membership. The Institute publishes codes of practice, advice notes, technical publications and, monthly, the journal *Wastes Management.*

Institute of Water and Environmental Management

See *Chartered Institute of Water and Environmental Management (CIWEM).*

insurance

The provision (usually by contract) of a legal obligation on the part of the person assuming the obligation (the insurer) to secure another person (the insured) against pecuniary loss falling upon the insured and arising out of one or more contingencies specified by the parties (incurred risks or perils insured against), in return for payment (the premium) paid by the insured to the insurer.

A contract of insurance (the policy) may be of two main types: occurrence policies, under which the insured is covered in respect of damage or loss arising out of the insured perils, whenever they arise (even although the damage in respect of which the claim is made manifests itself long after the policy is discontinued); and claims-made policies (under which the insured must make his or her claim within the period stipulated for that purpose in the policy, which may be as short as the policy year itself). Environmental risks, although covered under the broad wording of many general commercial liability policies, are now usually excluded from most insurance policies. Long-tail pollution liabilities under policies written as long ago as the 1950s contributed significantly to the Lloyds' capacity crisis of the early 1990s. Some policies still include cover for pollution which arises from sudden and accidental discharges, but liability for gradual pollution is difficult to obtain.

Some environmental impairment liability (EIL) *(qv)* policies have come onto the market recently, particularly policies aimed at possible liability for contaminated land, but the overall sums insured are low, the premiums high and the cover is always conditional on an environmental audit often involving intrusive investigation such as soil-sampling. All modern EIL policies are written on a claims-made basis.

There are a few cases where there is a statutory requirement for compulsory insurance against environmental risks, but this is presently confined to sectors which are in some senses untypical. Examples include marine carriage of oil, where the cover is absorbed by specialist, mutual insurance societies; and nuclear liabilities where there is a very high degree of governmental involvement.

integrated gasification and combined cycle

A clean combustion process for the combustion of liquid and solid fuels which may now constitute the best environmental option *(qv)*, but which may be more expensive than flue gas desulphurisation (FGD) *(qv)*.

integrated pollution control (IPC)

A system of control over industrial, environmental emissions which reflects the interdependence of the environmental media (water, atmosphere and land) by considering all together, and issuing a single authorisation. IPC was introduced in England and Wales by the Environmental Protection Act 1990, Part I, and applies to the prescribed processes set out in the Environmental Protection (Prescribed Processes and Substances) Regulations 1991, but only if those processes involve a release into the environment of one of the prescribed substances also specified in the Regulations. These substances differ for each of the receiving media (eg those for water are those on the red list *(qv)*). If a process does discharge a prescribed substance, all discharges (including those of non-prescribed substances) are regulated under the IPC process. The process is administered by Her Majesty's Inspectorate of Pollution (HMIP) *(qv)* and a parallel system operates for local air pollution controls *(qv)* administered by local authorities. Two key concepts underlie IPC: the obligation on operators to use the best available techniques not entailing excessive cost (BATNEEC) *(qv)*, and the duty on HMIP to ensure that the methods adopted in control of the activity represent the best practicable environmental option (BPEO) *(qv)*. Despite its breadth, IPC is not fully integrated as, for example, it does not regulate the final disposal of waste on land. The IPC system is being phased in over a period of several years and will eventually apply to all plants carrying on prescribed processes, whether new or existing at the date of the introduction of IPC controls.

The system applies in Scotland (but the timetable differs from that in England and Wales), where its administration is largely undertaken by Her Majesty's Industrial Pollution Inspectorate for Scotland *(qv)* with some functions discharged by the river purification boards *(qv)*. IPC does not yet apply in Northern Ireland.

integrated pollution prevention and control (IPPC)

IPPC is a variant of the United Kingdom's system of integrated pollution control *(qv)*, and it is currently under consideration, in the form of a draft directive, in the European Communities (EC) with the ambition of applying it to existing processes by July 2000 (and earlier for new processes). In a significant departure from the United

Kingdom model upon which it is based, the draft directive proposes the criterion of best available technology (BAT) *(qv)* instead of best available techniques not entailing excessive cost (BATNEEC) *(qv)*.

Interdepartmental Committee on the Redevelopment of Contaminated Land (ICRCL)

A committee of members of Government Departments which was set up in 1976 to develop and co-ordinate guidance on contaminated land *(qv)*. It published a series of Guidance Notes for local authorities, dealing with the redevelopment of landfill sites, gasworks, sewage works and scrapyards; and also giving general advice on asbestos, fire hazards and assessing contaminated land. These remain useful sources of information, although the House of Commons Environment Committee was highly critical of the Committee's performance and thought that its interdepartmental character was no longer of great significance (*Contaminated Land* (1989–90; HC170). They urged that it should be abolished and that central responsibility for contaminated land should be assumed by the Department of the Environment.

International Commission on Radiological Protection (ICRP)

An independent non-governmental body of scientific experts drawn from round the world. It was established in 1928 with the name of the International X-ray and Radium Protection Committee, and restructured and renamed in 1950 in order to cover more effectively the rapidly expanding field of radiation protection. The ICRP's function is to provide guidance, in the form of published recommendations, on the fundamental principles upon which appropriate radiation protection measures can be based. In doing so it draws on current scientific and medical knowledge and on the work of other distinguished bodies, in particular the United Nations Scientific Committee on the Effects of Atomic Radiation (UNSCEAR). It is intended that detailed guidance on the application of its recommendations, either in regulations or codes of practice, should be elaborated by the various international and national bodies that are familiar with what is best for their needs, such as the United Kingdom's National Radiological Protection Board.

International Convention on Civil Liability for Oil Pollution Damage 1969 (CLC)

Colloquially known as the Civil Liability Convention (or CLC), the Convention was negotiated in the aftermath of the *Torrey Canyon* tanker wreck off the Cornish coast in 1967. In its original form, the Convention imposed strict liability on the owner of a laden tanker which caused oil pollution damage resulting from a spill of persistent oil (including crude oil, fuel oil and heavy diesel oil). In 1984, an attempt was made to update the CLC by a Protocol which would have raised the maximum sum recoverable from the ship-owner. The Protocol also sought (among other things) to include compensation for preventive measures (recoverable under the Tanker Owners' Voluntary Agreement concerning Liability for Oil Pollution (TOVALOP) *(qv)*, but not under the CLC) and for the restoration of the environment in the definition of 'pollution damage' and to allow recovery for damage caused in the exclusive economic zone. The 1984 Protocol was calculated on the assumption that the United States (which had not joined the CLC because it considered that its provisions were not stringent enough) would ratify them, but this expectation was disappointed, so in 1992 a further Protocol was negotiated, reproducing many of the terms of the 1984 text but with more modest entry into force provisions. The Convention is administered by the International Maritime Organisation *(qv)*.

International Convention on the Establishment of an International Fund for Compensation for Oil Pollution Damage 1971

Otherwise known as the Fund Convention, and the partner convention to the International Convention on Civil Liability for Oil Pollution Damage (CLC) *(qv)*. The Fund Convention represents the contribution of oil cargo interests (as opposed to ship-owners) towards compensating victims of oil pollution damage, by raising the maximum sum recoverable. In its original form, the Fund Convention also assumed the top slice of the ship's liability, which was repaid to the owner under the roll-back provisions. Along with the CLC, it is supported by two voluntary agreements, the Tanker Owners' Voluntary Agreement Concerning Liability for Oil Pollution (TOVALOP) *(qv)* and the Contract Regarding a Supplement to Tanker Liability for Oil Pollution (CRISTAL) *(qv)*, which apply where there are uncompensated claims under the Fund.

A Protocol to the Fund Convention was negotiated in 1984, to complement the revision of the CLC, but did not come into force. A further Protocol was agreed in 1992 which broadens its scope and enhances the compensation provisions, but is not yet in force. The Fund established by the Convention is an international legal person, based in the headquarters of the International Maritime Organisation *(qv)* and it is financed by a levy on oil imported by sea into Member States.

International Convention on the Prevention of Pollution from Ships 1974 (MARPOL)

This Convention was negotiated to replace the London Convention on the Pollution of the Sea by Oil 1954–69 (OILPOL), and it is now the most far-reaching international convention dealing with vessel-source pollution. MARPOL contains the most widely accepted code of international rules relating to the construction, certification, inspection and operation of vessels carrying environmentally dangerous substances by sea. It is a framework convention, the provisions of which are amplified by five annexes, dealing with the carriage of oil and chemicals and with the discharge of sewage and garbage from ships.

international law

A corpus of binding legal obligations recognised by States and by international organisations. International law is generally regarded as being derived from treaties, customary principles of sufficient certainty and uniformity recognised over a sufficient period by civilised States, the decisions of the International Court of Justice and the writings of distinguished scholars in the subject.

International Maritime Organisation (IMO)

A specialised agency of the United Nations, known until 1982 as the Intergovernmental Maritime Consultative Organisation (IMCO). IMO has responsibility for a number of conventions which are concerned with the protection of the marine environment, including the Safety of Life at Sea Convention, International Convention on Civil Liability for Oil Pollution Damage, the International Convention on the Establishment of an International Fund for Compensation for Oil Pollution Damage, the International Convention on the Prevention of Pollution from Ships, the International Convention on Standards of Training, Certification and Watchkeeping for Seafarers, the International Convention on Maritime Search and Rescue and the International Convention on Oil Pollution Preparedness, Response and Co-operation. Many of these Conventions have been developed by IMO's Legal Committee. Other IMO Committees have responsibility for the protection of the marine environment and maritime safety issues. IMO runs an active technical assistance programme for developing countries.

International Organisation for Standardisation (ISO)

A worldwide federation of national standards bodies (the British Standards Institution in the UK) which officially began to function on February 23, 1947. The object of ISO (the short form of the Organisation's name is ISO in all languages, being derived from the Greek *isos* meaning 'equal') is to promote the development of standardisation and related activities in the world with a view to facilitating international exchange of goods and services, and to developing co-operation in the spheres of intellectual, scientific, technological and economic activity. The work of ISO brings together the interests of producers, users (including consumers), governments and the scientific community and is carried out through 2,754 technical bodies with the assistance of more than 30,000 experts from all parts of the world. To date, over 9,000 ISO standards have been published, including ISO 14001 (still awaiting trial approval) on Environmental Management Systems Specification with guidance for use; and ISO 14015 on general principles of Environmental Auditing and Related Environmental Investigations.

International Solid Waste Association (ISWA)

The international equivalent of the Institute of Wastes Management (IWM) *(qv)* with the objective of promoting or developing professional solid waste management world-wide amongst some 30 member countries. Solid waste management in its broader sense is interpreted by ISWA to mean integrated systems for waste reduction, collection, transport, recycling, energy recovery treatment and disposal. ISWA is presently carrying out a review of the capabilities of national members to carry out their function which displays an unusually healthy self-introspection. [*Address: International Solid Waste Association (ISWA), General Secretariat, Bremerholm 1, DK–1069 Copenhagen K, Denmark; T: +45 33 91 44 91; F +45 33 91 91 88*].

International Tanker Owners' Pollution Federation (ITOPF)

An organisation founded in October 1969 by the oil tanker industry in the aftermath of the *Torrey Canyon* and other tanker disasters in order to administer the Tanker Owners' Voluntary Agreement concerning Liability for Oil Pollution (TOVALOP) *(qv)*. The Agreement had been established to provide interim cover, pending the entry in force of the International Convention on Civil Liability for Oil Pollution Damage (CLC) *(qv)*. In addition to servicing the TOVALOP agreement, ITOPF provides technical services and advice on clean-up operations world-wide, assists in the assessment of claims for compensation for oil pollution damage, and in the preparation of contingency plans for oil pollution incidents. It also conducts an active training and education programme.

International Union for the Conservation of Nature and Natural Resources (IUCN)

An international conservation organisation with headquarters in Gland, Switzerland. It has since changed its name to the World Conservation Union, but retains the old acronym IUCN. Its distinctive feature is its membership structure, which consists of States, governmental agencies and non-governmental organisations. Largely a scientific organisation, with extensive field projects engaged in resource management, conservation and sustainable development programmes, IUCN can also call upon the experience and expertise of hundreds of individual specialists, who are enrolled in Commissions, among which is the Commission on Environmental Law. That Commission shares with the IUCN Environmental Law Centre responsibility for executing the legal elements of the IUCN programme (approved by a General Assembly held every three years). The Commission is particularly concerned with the development of the doctrinal principles of environmental law, both at the national and international level. The Environmental Law Centre maintains an extensive databank of environmental legislation and supports the Commission in its activities. Both the Commission and the Environmental Law Centre are located in Bonn, Germany.

International Union of Pure and Applied Chemistry

The international scientific union of national organisations of chemists. Its aims are the promotion of continuing co-operation among chemists of member countries; the working out of recommendations for international regulation, standardisation or codification of areas of chemistry; co-operation with other international organisations which deal with topics of a chemical nature; and promotion of progress in the field of chemistry. It was founded in 1919 in Paris, to carry on the work of the International Association of Chemistry Societies.

inversion

A climatic phenomenon which often increases atmospheric pollution. Refers to temperature inversion in the atmosphere, in which the temperature, instead of falling, increases with height above the ground. With the colder and heavier air below upward currents do not form and turbulence is suppressed. Inversion often occurs in the late afternoon when the radiation emitted from the ground exceeds that received from the sinking sun. It is also caused by katabatic winds, which are cold winds flowing down the hillside into a valley, and by anticyclones.

In inversion layers, both vertical and horizontal diffusion is inhibited and pollutants become trapped, sometimes for long periods. Low-level discharges of pollutants are more readily trapped by inversions than high-level discharges; furthermore, high-level discharges into an inversion tend to remain at a high level because of the absence of vertical mixing.

ion

An electrically charged atom or group of atoms.

ion exchange

A process in which ions *(qv)* in solution are exchanged with ions from a solid material, usually a synthetic resin.

irrigation

The artificial collection, storage and distribution of water onto land. Also a form of sewage disposal involving spraying effluent onto land.

ISO 14001

An environmental management system developed by the International Organisation for Standardisation *(qv)*.

joint and several liability

The liability of more than one defendant, normally in a civil action. Their liability is joint when each is liable for his or her own contribution to the single act or item of damage caused to the plaintiff; but is several when each is also liable to the plaintiff for damage caused by the other, though with a right to recover a contribution from the other defendant(s). Hence it simplifies litigation for the plaintiff, especially where it is unclear which of a class of defendant contributed in which way to the damage, and where one or more of the defendants is untraceable or insolvent. Joint and several liability is a feature of the US Superfund *(qv)* scheme of liability for contribution to the clean-up costs of contaminated sites, and has enabled the US Environment Protection Agency to concentrate its action on those defendants best able to foot the bill (so-called 'deep pocket' liability). There is a wide class of potential contributors, called potentially responsible parties, who under certain circumstances must share joint and several liability for the total damage.

Joint Nature Conservation Committee

The statutory joint committee of English Nature *(qv)*, Scottish Natural Heritage *(qv)* and the Countryside Council for Wales, with responsibility under the Environmental Protection Act 1990 for the so-called 'special functions' specified in s.133, which include nature conservation issues affecting Great Britain as a whole or outside Great Britain. [*Address: Monkstone House, City Road, Peterborough, Cambs, PE1 1JY; T: 01733 62626.*]

joule

See *British thermal unit.*

judicial review

An administrative law procedure by which persons having sufficient legal standing *(locus standi) (qv)* may challenge in court the decision of a public authority. In the environmental field, the decision to be challenged is usually that of a Minister, a regulatory body (such as Her Majesty's Inspectorate of Pollution (HMIP) or the National Rivers Authority (NRA)) or a local authority *(qv)*, although following the judgment in *Griffin v South West Water Services* (as yet unreported) the operational decisions of bodies (typically utilities) in the private sector exercising public duties pursuant to special statutory powers may also be susceptible to judicial review.

The challenge may be made on one of a number of grounds, including error of law, the denial of the rules of natural justice, procedural irregularity or arbitrariness. The court has no power to review the decision on its merits, but if satisfied of the applicant's case, may quash the decision complained of, leaving it open to the decision-maker to come to a fresh decision in accordance with the law. The process is commenced by an application to the court for leave to apply for judicial review which must be made promptly, and in any case within three months of the act or omission complained of. If leave is granted, the substantive application may proceed. Judicial review is assuming increased importance in British environmental law because of the recent relaxation in the rules relating to standing (see *locus standi*), and because of the importance of EC environmental legislation which cuts across the broad administrative discretion traditionally conferred on public officials by United Kingdom environmental legislation.

Karin B

Perhaps the most notorious of the 1980s itinerant ships carrying hazardous waste. The *Karin B* was a German-flagged vessel which loaded hazardous waste in an attempt unlawfully to dispose of it in Nigeria. She unsuccessfully sought entry to the ports of a number of states, including the United Kingdom, before finally returning the waste to Italy, its state of origin.

knowingly permit

See *strict liability (criminal)*.

laminar flow

The smooth flowing of liquid or gas, as opposed to a turbulent flow.

land drainage

A term commonly applied to the system of flood protection presently the responsibility of the National Rivers Authority *(qv)* through regional flood defence *(qv)* committees under the Land Drainage Act 1991 and the Water Resources Act 1991. In some parts of the country which are particularly susceptible to flooding, internal drainage boards are established to manage watercourses and drainage works. These drainage bodies are responsible for the construction of new drainage works (which will usually require planning permission *(qv)*), maintaining or improving existing works and ensuring the free flow of watercourses. The Authority and the internal drainage boards have limited legislative powers to make byelaws, relating to their functions. See also *coastal protection*; *drainage*.

landfill

A method of waste disposal involving the burial of waste in the ground, frequently in voids resulting from the extraction of minerals. Conditions attached to waste management licences *(qv)* regulate the kinds of waste which each landfill site may accept. This is not only to ensure that special wastes *(qv)* are only landfilled (if at all) at suitable sites, but also to control the generation of leachate *(qv)* and landfill gas *(qv)*. A draft directive on landfill presently under consideration in the European Communities (EC) seeks to allow landfills to receive only certain stipulated types of waste and to restrict the UK practice of co-disposal *(qv)*. The draft directive also seeks to raise the standards to which landfills are to be operated throughout the EC. The EC has in the past expressed the opinion that the cost of disposal of waste to landfill is too low compared to other methods of disposal and should be increased. The UK Finance Act 1995 introduces a levy on landfill (see *landfill levy*). The EC regards landfill as the last resort in the hierarchy of waste management options, and landfill capacity in the United Kingdom is steadily diminishing. A report in 1994 from the South East Regional Planning Conference (SERPLAN) suggests that South East England will run out of landfill space within 10 to 15 years at current rates of disposal.

landfill gas

The complex mixture of dangerous gases formed during the decomposition of biodegradable wastes in a landfill site. Landfill gas comprises about 64% methane and 34% carbon dioxide, plus trace concentrations of a range of organic gases and vapours. Problems occur when landfill gases escape along paths of least resistance, moving within the landfill site and beyond the site boundary. The harmful effect of landfill gas lies primarily in its ability to displace oxygen from the soil, which is then unable to sustain plant growth. Other dangers associated with landfill gases are the risks of explosion, fires, asphyxiation and health hazards.

landfill levy

Announced in the Budget of November 1994 as a tax intended to deter landfilling of waste, encourage recycling and to raise revenue from waste disposal to help finance a cut in employers' National Insurance contributions. The rate of tax is open for discussion under the March 1995 Consultation Paper, but the Government is considering rates of 30% to 50% with VAT on the levied total. The estimates of the volume of waste disposed of annually are 100m tonnes, incurring disposal charges of about £1 bn, and therefore future Government revenue of between £300m and £500m.

The proposal to levy the tax on price, rather than volume, might well penalise expensive landfill operators which have invested in costly environmental engineering, and that form of tax would magnify the cost differences between cheap and cheerful waste disposal services, favouring the cheaper landfill operator which tended to have the lowest environmental standards. Greenpeace *(qv)* has also voiced fears that the focus on landfills will encourage waste producers to turn to incineration with potentially more damaging environmental consequences. A DOE 1993 report concluded that a 50% tax on a tonne of waste, ie £10 average disposal cost, would not promote recycling but would divert an extra 5% of domestic waste to incinerators.

Lappel Bank

An area of mudflat on the Isle of Sheppey in the Medway Estuary in Kent, alongside the deepwater port of Sheerness, where there has been conflict between the desire of the port to expand and the importance of the area as a nesting and breeding area for migratory species of birds. It has twice been the subject of litigation in proceedings brought by the Royal Society for the Protection of Birds (RSPB) *(qv)*, the first in relation to a grant of planning permission *(qv)* by the local planning authority *(qv)* in 1990 for port expansion (*R v Swale Borough Council, ex parte RSPB*) and again in 1994 when the House of Lords rejected an argument that, in designating a special protection area (SPA) *(qv)* under the EC Birds Directive, the Secretary of State should have considered only the ornithological features of the area, and not the economic requirements of the area. The House of Lords agreed to refer the dispute to the European Court of Justice *(qv)*, but refused to order a stay in the undertaking of the reclamation works.

leachate

Water that has become contaminated by passing through materials, commonly those which have been deposited on or in the ground, as in landfill *(qv)* sites. The control of pollution at landfill sites depends primarily on engineered management of leachate through containment, recirculation and other methods. Leachate is generally beneficial to the decomposition of materials in landfill sites. Past analysis of the contaminants contained in leachates from various landfill sites which have received domestic, as opposed to industrial, wastes has not shown any significant differences in their polluting qualities.

lead in petrol

Tetraethyl is a controversial process chemical added to petrol to increase its octane rating. As such, it is discharged in aerosolised form from vehicle emissions. It has been estimated that in 1986 some 300 tonnes per annum of lead were released in this form in the US (Hugh D Crone, *Chemicals and Society*, Cambridge University Press, 1986, p.135.) The policy approach in the UK and EC has been to control the lead content of vehicle fuel, particularly following the 1981 reform report of Professor P J Lawther, *Lead and Health* (HMSO), and the 1983 report of the Royal Commission on Environmental Pollution on Lead in the Environment (9th Report, Cmnd 8852).

The Government's response to the Lawther Report was to limit emissions of lead by lowering the content of petrol to 0·15 g/l. Following the 1983 Royal Commission Report, the Government announced that all new petrol engine vehicles should be constructed to run on lead-free petrol. The use of lead-free fuel was subsequently encouraged by tax differentials in the March 1987 and subsequent Budgets. The relevant EC legislation is Directive 85/210 on the approximation of laws concerning leaded petrol. The UK regulations are the Motor Fuel Lead Content of Petrol Regulations 1985, No 1728, consolidating SI 1976/1866 as amended by SI 1979/1 and 1981/1523. The Royal Commission on Environmental Pollution in its 18th Report on *Transport* (Cm 2674, 1984) noted the sharp decline in emissions of lead from vehicles as a result of these policies (para. 8.19).

The House of Commons 1994 Transport Committee's Report on Air Pollution suggested the expensive campaign to encourage motorists to switch to lead-free petrol has not just been a waste of effort and money but actually made pollution worse. The strongly worded report claims that lead-free petrol, sold from green pumps to advertise its environmental credentials, causes more cancer and childhood leukaemia and is more likely than leaded petrol to produce fumes that linger. The Committee called for an immediate ban on the sale of super unleaded petrol on the grounds that it is dangerous to use in cars without a catalytic converter. The petroleum industry strongly resists these conclusions and the debate continues.

Leaking Underground Storage Tank (LUST)
The seeping of substances held in underground storage tanks has become a matter of environmental concern because of the possible pollution of groundwater *(qv)*, particularly in the US, where the US EPA coined this acronym. Leakage is usually gradual as the tank corrodes. It can also be sudden if the corrosion results in seals or walls giving way. See also *underground storage tank.*

lender liability
The civil liability *(qv)* that may attach to those who have lent money to or otherwise financially supported a person who becomes responsible for environmental remediation costs, such as the clean-up of contaminated land. The doctrine goes beyond merely the fall in value of the lender's security when it is against the land found to be contaminated. Under certain conditions, a lender may become liable also to meet or contribute to the clean-up costs. Such cases include a lender who takes possession of the land in order to realise his security, and may then be found to be an 'owner' for liability purposes, or even the operator of the contaminating facility; or where a lender has participated in some direct way in the management of the facility, such as by having a nominee on the board of directors. The leading case is a decision of the US Federal Court in *United States v Fleet Factors Corporation* (1990) 901 F.2d 1550 (11th Cir. 1990), which suggested that, in certain circumstances, lenders might become directly liable for environmental harm caused by their borrowers' operations. How far such an approach should be encouraged in Europe was an issue raised by the European Commission's 1993 Green Paper on environmental liability *(qv)*. The Environment Bill 1995 *(qv)* contains no specific provision to exempt lenders, hence allowing lender liability to arise wherever a lender's proximity to the site or the borrower's business activities renders him either an 'owner' or 'occupier' of the land, or a person who has caused or knowingly permitted the contaminating substances to be deposited there.

less sensitive waters
A classification under EC Directive 91/271 on urban waste water treatment, also known as high natural dispersion areas, in relation to estuaries and coastal waters where discharges from sewage treatment plant will not have an adverse effect on the environment, and where

therefore under the Directive only primary treatment will be required so as to bring about a percentage reduction in load of biochemical oxygen demand (BOD) *(qv)* and suspended solids. Primary treatment involves the physical treatment of sewage effluent, usually by settlement, to remove gross solids and reduce suspended solids by about 50%, and BOD by about 20%. See also *normal waters; OFWAT; sensitive waters; Urban Waste Water Treatment Directive (UWWTD).*

lethal concentration
The amount of a given substance which will, by some means through whatever application, cause the death of one or more living organisms.

lethal dose (LD50)
A statistically derived single dose of a substance that can be expected to cause death in 50% of animals.

lifecycle analysis (LCA)
A process to evaluate the environmental burdens associated with a product, process or activity. An inventory of data on resource consumption, pollutant releases and waste generation associated with the relevant product, process or activity is compiled, the impact of those energy and material uses and releases into the surrounding environment is assessed, and opportunities for environmental improvement are identified and evaluated. There is currently no internationally accepted methodology for this process. However, the Society of Environmental Toxicology and Chemistry in Brussels has developed guidelines aimed at encouraging transparency in this area, and these are increasingly being accepted as an authoritative international code.

limestone pavement
An area of limestone that has been fissured by natural erosion and is of physiographical, and potential biological, interest. It may be protected by the making of a limestone pavement order under the Wildlife and Countryside Act 1981, s.34 and Sched 11, where it is considered to be of special interest. A limestone pavement order prohibits the removal or disturbance of limestone from the designated area. Limestone pavements are identified as a priority habitat type under the EC Habitats Directive (91/43).

limit values
As used in environmental legislation of the European Communities (EC) *(qv)*, an outer limit within which Member States must set environmental quality standards *(qv)* and emission standards *(qv)*. Usually, the limit is expressed in numerical form defining the

concentration of the pollutant in question which can lawfully be permitted in the discharge being regulated, eg x micrograms of pollutant per litre of effluent discharged. Limit values may be implemented in Member States by using emission standards, often uniform emission standards *(qv)*, which is the preference of most Member States or, in some cases, by using environmental quality standards *(qv)*, which has been a preferred approach in the United Kingdom.

lindane

This is also commonly referred to by its chemical names of gamma benzene hexachloride (BHC) and gamma hexachlorocyclohexane (HCH). An insecticide and herbicide consisting of a white, water-insoluble powder, $C_6\,H_6\,C_6$.

L'Instrument Financier pour l'Environnement (LIFE)

The European Communities' (EC) Financial Instrument for the Environment (LIFE), which was adopted by the Council of Ministers on May 21, 1992 (Regulation No 1973/92) and came into force on July 23, 1992. The LIFE programme is the main source of EC funding for environmental initiatives with emphasis upon projects with European interest. Its objective is to contribute to the development and implementation of EC environmental policy and legislation. The scope of LIFE is defined by five fields of action:

1 the promotion of sustainable development and the quality of the environment;

2 the protection of habitats and nature;

3 administrative structure and the environment services;

4 education, training and information; and

5 actions outside the Communities' territory.

The LIFE programme is administered by the European Commission, assisted by a Management Committee made up of representatives from each Member State.

liquefied natural gas (LNG)

Natural gas which has been liquefied, usually for the purpose of storage, handling or transport, by a combination of compression and refrigeration. LNG has a very low flash-point and is potentially very explosive.

liquefied petroleum gas (LPG)

Propane or butane (or a mixture of the two) which has been liquefied, usually for the purpose of storage, handling or transport.

liquidated damages

A genuine and justifiable pre-estimate agreed between contracting parties of the loss which is likely to flow from a breach of contract and which is recoverable from the defaulting party without the necessity of proving the actual loss suffered. The term is also applied to sums expressly made payable as liquidated damages under statute. See also *penalties.*

listed building

A building of special architectural or historic interest recorded in the statutory list under the Planning (Listed Buildings and Conservation Areas) Act 1990, s.1 (in Scotland, the Town and Country Planning (Scotland) Act 1972, s.52). It is an offence to carry out any works to a listed building which would affect its character as a building of special architectural or historic interest, except in accordance with a special consent (listed building consent) for which application must be made to the local planning authority *(qv).* Although the legislation falls short of imposing a positive obligation on owners to maintain listed buildings, a local planning authority has power to carry out urgent repairs to unoccupied buildings and parts of buildings and charge the cost back to owner; and to acquire compulsorily any listed building in a state of repair if the owner fails to carry out repairs specified by the local planning authority. Further reference: PPG15, *Planning and the Historic Environment* (1994).

lithosphere

The Earth's crust, enclosing the kernel of the Earth or barysphere, extending to a depth of about 80km from the Earth's surface. Only the outer part, the biogeosphere, is associated with any form of life.

litigation

Strictly, proceedings in court in pursuance of a writ or summons, but widely used to refer to all contentious matters connected with the resolution of a dispute by court proceedings, arbitration or alternative dispute resolution. Sir Frederick Pollock once defined litigation as 'a game in which the court is the umpire'. Most experienced litigants would agree that the law, like the Ritz Hotel, is open to everybody.

litter

Items of discarded rubbish, especially when scattered in a public place.

It is an offence under the Environmental Protection Act 1990, s.87, to throw down, drop or otherwise deposit and leave 'any thing whatsoever in such circumstances as to cause, or contribute to, or tend to lead to, the defacement by litter of any place' to which the section applies. There is authority, under the now-repealed Litter Act 1958, for the proposition that this offence consists of two elements, the deposit of the litter and the failure to remove it, both of which must be proved to secure a conviction and further that the deposit is an event 'fixed in time' and cannot therefore constitute a continuing offence (*Vaughan v Biggs* [1960] 2 All ER 473, per Lord Parker C J at p.474).

loam

A rich friable soil containing more or less equal amounts of sand, silt, clay and usually organic matter.

Local Agenda 21

An initiative developed by local government in the United Kingdom under chapter 28 of Agenda 21 *(qv)*, which encourages local authorities to adopt a Local Agenda 21 for their communities by 1996. Local authorities are effectively beginning to define their own sustainable development *(qv)* strategies at local level, and to develop local environmental initiatives, creating partnerships with the business and voluntary sectors. Further reference: Local Government Management Board, *Agenda 21: A Guide for Local Authorities in the UK* (1993).

local authority

One of the elected councils responsible for local government in the United Kingdom. In most parts of England and Wales, there are two tiers of local government: the county councils and the district, borough or city councils. In Scotland, the former are replaced by regional councils and by three unitary islands councils. In London and metropolitan areas, there are unitary authorities (32 London borough councils and 36 metropolitan borough councils). From April 1, 1996 the pattern in Wales will change to 22 unitary councils (Local Government (Wales) Act 1994), and in Scotland to 29 unitary councils plus three islands councils (Local Government (Scotland) Act 1994). In England, any change to a unitary local government structure looks likely to be more pragmatic and patchwork, in accordance with local circumstances and wishes, as a result of the recommendations of the Local Government Commission for England under the Local Government Act 1992.

local authority air pollution control (LAAPC)

A system of control over atmospheric pollution introduced by Part I of the Environmental Protection Act 1990 in conjunction with the institution of integrated pollution control (IPC) *(qv)*. The processes which are subject to LAAPC are those which are intrinsically less polluting than IPC processes, either because of their scale or their nature. These processes are listed in Part B of Schedule 1 to the Environmental Protection (Prescribed Processes and Substances) Regulations 1991 (SI 1991 No 472) and are therefore colloquially referred to as Part B processes.

local authority waste disposal company (LAWDC)
An arm's length company to which a local authority *(qv)* has transferred its former undertaking as a waste disposal authority *(qv)* under the provisions of the Environmental Protection Act 1990, s.32 and Schedule 2. Local authorities had previously combined the role of waste disposal operator with that of the enforcement agency charged with the implementation of the law relating to waste management. Under the 1990 Act, the local authority's functions are confined to regulation and their former waste management and disposal operations were transferred either to a waste contractor in the private sector or to a LAWDC. The LAWDC itself need not remain in local authority hands, but may be privatised completely. So long as it is not privatised, however, its management must be distanced from local authority control as required for arm's length companies under the Local Government and Housing Act 1989, s.68. This means that directors must have fixed terms of service, and that not more than one-fifth of the board can be members or officers of the local authority.

local nature reserve
An area declared by a local authority under the National Parks and Access to the Countryside Act 1949, s. 21. The local authority has power to make by-laws for the protection of a local nature reserve.

local planning authority
The local authority *(qv)* to which is allocated functions under planning legislation. Different authorities may be the local planning authority for different purposes, or concurrently, in respect of the same area. For most purposes the functions of a local planning authority are allocated to the district council (or London borough), but in areas where there is still a two-tier structure of local government (see *local authority*) the county council (or, in Scotland, the regional council) is also the local planning authority for certain functions, including strategic planning through the preparation of structure plans, and development control *(qv)* for minerals and (not in Wales) waste. In urban development areas, the function of local planning authority in relation to development control and related

functions is allocated (except in Cardiff) to the urban development corporation, but remains with the district and county councils in relation to the preparation of development plans *(qv)*.

In national parks, the planning authority takes the form of a joint or special planning board, made up of representatives of the local planning authorities over whose areas the park extends. In Scotland, planning functions are distributed between the regional and district councils (except in the Border, Dumfries and Galloway, the Highlands and the islands councils, where there is a unitary system for planning purposes). In Northern Ireland, planning functions are discharged by the Department of the Environment for Northern Ireland.

locus standi

The interest which an applicant for judicial review *(qv)* must demonstrate that he or she has in the matter in dispute. The requirement has traditionally acted as a filter to prevent proceedings being instituted by busybodies with no real interest in the issues being raised, but has sometimes precluded legal challenge to official decisions from being made by environmental and other interest groups. This approach is epitomised by the decision in *R v Secretary of State for the Environment, ex parte Rose Theatre Trust Company* [1990] 1 PLR 39; [1990] 1 All ER 754; [1990] JPL 360 that a company created by a group of archaeologists and actors had no *locus* to challenge the refusal by the Secretary of State to schedule the site of Shakepeare's Rose Theatre as an ancient monument. Environmental legislation in the United Kingdom establishes no special rights for environmental groups to commence proceedings. However, the courts have now adopted a more accommodating approach, and have upheld the *locus standi* of the British Herpetological Association in relation to a site where it had undertaken research (*R v Poole Borough Council, ex parte Beebee* [1991] 2 PLR 27). Most significantly, they have now come to recognise the advantages of allowing *locus* to environmental organisations with a proven track-record of involvement in the environmental issues concerned in the case and with specialist technical or other knowledge to contribute to the resolution of the issues to be determined. In principle, such organisations have *locus*, at least where they have individual members who are directly affected by the decision being challenged: see *R v HMIP, ex parte Greenpeace* [1994] 4 All ER 321.

London Waste Regulatory Authority

A statutory joint authority which acts as the waste management authority for the area of the former Greater London Council, from which it inherited its powers. It is due to be abolished under the Environment Bill 1995, and absorbed into the new Environment Agency *(qv)*.

Lugano Convention

The Convention on Civil Liability for Damage Resulting from Activities Dangerous to the Environment, concluded in Lugano in 1993 under the auspices of the Council of Europe *(qv)*. It imposes strict liability on those who cause pollution of the environment. It has not been ratified by the United Kingdom, and for Member States of the European Communities (EC) looks likely to be overtaken by a Directive on environmental liability, following further consultation on the Commission's 1993 Green Paper: see also *environmental liability*.

Maastricht

The Dutch town that gave its name to the Treaty on European Union, signed there on February 7, 1992, which made several fundamental amendments to the Treaty of Rome. The Maastricht amendments were integrated with United Kingdom national legislation by the European Communities (Amendment) Act 1993. The Treaty amended *inter alia* the environmental provisions of the principal treaty, which had been introduced by the Single European Act in 1986, principally by amending the procedural bases for environmental legislation and enabling the Community to act (but only by unanimity) in the areas of town and country planning and fiscal measures).

mad hatter's disease

A disease suffered by hatters of a century ago, who became mentally deranged through absorbing mercury used in making felt for hats. Not thought to be particularly prevalent in contemporary official circles today.

main river

That part of the river system which is defined on a statutory main river map held by the Ministry of Agriculture, Fisheries and Food, the Welsh Office and the National Rivers Authority. All other rivers are classified as ordinary watercourses or surface water sewers.

management agreement

An agreement securing the conservation and management of a site possessing particular qualities in relation to nature or countryside objectives, usually a nature reserve or a site of special scientific interest (SSSI) *(qv)*. Management agreements are made under the National Parks and Access to the Countryside Act 1949, s.16, the Countryside Act 1968, s.15, the Countryside (Scotland) Act 1967, s.49A and the Conservation (Natural Habitats, &c.) Regulations 1994 (SI 1994 No 2716), reg. 16 (in relation to land in or adjacent to a European Site *(qv)*) between an owner, lessee or occupier of land and an authority exercising a conservation function in respect of the land, and (unless they provide otherwise) are binding upon subsequent owners of the land. Depending on the statutory power involved, the relevant authority may be a local authority, the Nature Conservancy Council for England *(qv)*, the Countryside Council for Wales *(qv)* or Scottish Natural Heritage *(qv)*. The object of the agreement is commonly to ensure the management of land in order to provide suitable conditions for the study and preservation of flora and fauna and geological and physiographical features and for SSSIs, and to conserve or enhance the natural beauty or amenity of such land. Restrictions contained in such agreements may relate to agricultural practices or the carrying out of building works and positive obligations may permit the authority to carry out necessary operations on the land. Appropriate compensatory payments are made to reflect the diminution in the value of land to the person with the interest in it.

mangroves

Plant communities and trees that inhabit tidal swamps, muddy silt and sand banks at the mouths of rivers and other low-lying areas which are regularly inundated by the sea, but which are protected from strong waves and currents. Mangroves are the only woody species which will grow where the land is periodically flooded with sea water; individual species have adapted themselves to different tidal levels, to various degrees of salinity and to the nature of the mud or soil. Mangroves vary in size from substantial trees up to 30 metres in height down to miniature forms less than waist high. Mangrove swamps and thickets support hundreds of terrestrial, marine and amphibian species.

marine nature reserve (MNR)

A designation introduced by the Wildlife and Countryside Act 1981, s.36, for any area of land covered (continuously or intermittently) by tidal waters, or to parts of the sea up to the seaward limits of territorial waters. The objective of designation is to conserve marine flora or fauna or geological or physiographical features, or to allow study of such features. MNRs are actively managed by the relevant conservancy councils who have the power to make by-laws to restrict killing, taking and disturbance of plants and animals, on the deposit of litter and to restrict access by people on pleasure boats. To date, only one designation has been made, at Lundy Island off the North Devon coast.

mass balance

A tabulation of all material entering and leaving a system. Under steady conditions, the total mass entering the system in unit time must equal the total mass leaving. In particular, the masses of each chemical effluent entering and leaving must balance. See also *effluent.*

mass transfer

A process in which matter passes in the molecular state across the boundary between two phases, as from a gas into a liquid.

material change in use

The making of any material change in the use of any building or other land constitutes development *(qv)* under the Town and Country Planning Act 1990, s.55, and is lawful only if:

1 it is undertaken in accordance with planning permission, or

2 it constitutes permitted development under the Town and Country Planning (General Permitted Development) Order 1995, or

3 the new use is in the same class as the former use under the Town and Country Planning (Use Classes) Order 1987.

The carrying out of building, engineering, mining or other operations also constitutes development requiring planning permission under the 1990 Act. A change in a manufacturing process does not of itself constitute development, and is outside planning control. Hence the need for supplementary controls in relation to the storage on land of hazardous materials under the Planning (Hazardous Substances) Act 1990.

maximum exposure limit (MEL)

A limit prescribed by the Health and Safety Commission *(qv)* for a particular substance. Control measures must then ensure that the level of exposure is reduced as far as is reasonably practicable and must be below the specified MEL. If an employee uses a substance which gives off a fume or vapour for which an MEL is specified, then the employer must take all necessary control measures to reduce the level of the fuel or vapour inhaled by the employees so that it is below the MEL. See also *Control of Substances Hazardous to Health Regulations 1994.*

mercaptans

Organic compounds generally produced in oil refinery feed preparation units as a result of incipient cracking. The simpler mercaptans have strong, repulsive, garlic-like odours, and have in the past often been the cause of neighbourhood complaint when deposited at landfill sites in an insensitive manner with no regard for wind direction and the proximity of community activity or housing.

mercury

A heavy liquid metal (also known as quicksilver) which is potentially highly toxic and in humans may induce irritability, paralysis, blindness, insanity, muscle tremors and birth defects. Mercury is used in older chlor-alkali *(qv)* plants, the pulp and paper industry, and in seed treatment, and is released to the environment in the burning of fossil fuels and mining and refining processes. Mercury compounds may function as an accumulative poison affecting the nervous system, particularly in the form of methyl mercury. The presence of mercury does not necessarily indicate a condition of pollution; mercury in the ocean, for example, is largely of natural origin. Inorganic salts of mercury are generally not a threat to living creatures and are widely used in agriculture to control disease and pests. In contrast, organic compounds used to control fungal diseases in seeds, growing plants, fruits and vegetables can be absorbed by humans and can be extremely toxic. The most notorious incident of mercury poisoning is that affecting the community of Minimata in Japan during the 1970s, the effects of which are still being experienced. This was the result of the accumulation of mercury through the prolonged ingestion of fish. Further reference: House of Lords Select Committee on the European Communities, *Water Pollution: Mercury* (1980).

mesophyte

A plant which requires an average amount of moisture. Most trees, shrubs and herbs living in climates of moderate rainfall and temperature are mesophytes.

methaemoglobinaemia

A disorder which can affect the amount of oxygen in the blood. Known as blue baby disease, it was thought to be associated with drinking water containing nitrogen.

methane

An inflammable hydrocarbon gas which is a constituent of landfill *(qv)* gas and at times liberated during sewage treatment. It is also the principal constituent of natural gas with a chemical formula of CH_4. It is a powerful greenhouse gas *(qv)*.

microbiology

The biology of microscopic organisms.

mine, abandoned

A mine, normally underground, at which mineral extraction has ceased without reasonable prospect of resumption. There is presently no statutory definition, although for reasons of working conditions and safety within mines, notice must, under the Mines and Quarries Act 1954, be given by a mine owner to the district inspector of mines of the abandonment of an entire mine, or a seam or vein within it.

The environmental consequence of abandoned mines arises from the fact that in working mines water is actively removed by pumping, and treatment facilities are installed to reduce the potentially harmful environmental impact of the sometimes highly contaminated waters. Because of the geology of coal mining areas, rising water is likely to come into contact with pyrites which have been exposed to the air due to the mining activities. The subsequent chemical reactions form sulphuric acid in the minewater. This acidity allows the water to carry high concentrations of iron, sulphate, aluminium and a range of heavy metals depending on the geology and strata through which the water will rise. The water may continue to rise until it reaches the level of a watercourse such as a river. The highly acidic metal-laden water running into the river may kill aquatic life for considerable distances. In addition, the dilution of the acidity causes sulphates and metals to precipitate out of the water, thereby coating the river bed in an orange-red blanket. It is difficult to predict to what level the water table will rebound. This has led to speculation that some previously dry areas could become prone to flooding or waterlogging and that buildings and roads could become unstable. Foundations may also be attacked by sulphate-rich water. Rising water may also displace dangerous mine gases, such as methane and carbon monoxide.

Once a mine is abandoned, an exemption arises under water pollution legislation. Although it is an offence under the Water Resources Act 1991, s.85, to cause or knowingly permit pollution to enter any controlled waters, there is an exception for a person who permits (but not the person who causes) water from an abandoned mine to enter controlled waters. Moreover, the power of the National Rivers Authority under s.161 of that Act to charge clean-up costs back to polluters is also excluded in the case of abandoned mines.

The Environment Bill 1995 *(qv)* proposes a new s.91A in the Water Resources Act 1991 which will provide an extended definition of 'abandonment' for these purposes, one head of which is the discontinuance of water pumping operations. The Bill also proposes that the present exemption should lapse in respect of any mine abandoned after December 31, 1999. In advance of coal privatisation, a memorandum of understanding was reached between the British Coal Corporation and the National Rivers Authority in November 1993, but it does not commit British Coal to preventing pollution from abandoned mines. Further reference: National Rivers Authority, *Abandoned Mines and the Water Environment* (Water Quality Series No 14; 1994).

mineral planning guidance (MPG)

Guidance published from time to time by the Secretary of State for the Environment to provide specific advice on the control of minerals development. The series started in 1988, and current MPGs contain guidance on opencast mining, aggregates, reclamation of mineral workings and control of noise at surface mineral workings. The policy contained in an MPG note is a material consideration for the purposes of planning control but it does not have the force of law. See also *planning policy guidance.*

mining

The extraction of minerals from the ground, which may be by surface working (opencasting) or underground working. Mining operations constitute development under the Town and Country Planning Act 1990 (where they are given a broad definition as including the removal of material of any description from a mineral working deposit, a deposit of pulverised fuel ash or other furnace ash or clinker or from a deposit of iron, steel or other metallic slag and the extraction of minerals from a disused railway embankment), and hence normally require planning permission. Mining dereliction has been a major inherited environmental problem, and legislation now allows conditions to be imposed when mineral planning permission is granted, requiring the restoration and aftercare of the site when mining ceases. See also *mine, abandoned.*

Ministry of Agriculture, Fisheries and Food (MAFF)

A Government Ministry with wide-ranging responsibilities in England and Wales for agriculture, which include some environmental controls. It issues Codes of Practice on matters such as fertilisers and pesticides, for which it is also the registration authority. MAFF monitors levels of radioactivity in water, soil and food and oversees the implementation of the European Communities' (EC) agricultural and fisheries policies and also licences.

The Land Use and Tenure Division is concerned with planning implications on agricultural land, farm woodlands and forestry and the Department of Agriculture is responsible for designating environmentally sensitive areas (ESAs) *(qv)*. MAFF also develops environmental protection policies on whaling, fish diseases, salmon and freshwater fisheries and has flood protection and sea defence functions. It also administers grant schemes to facilitate the improvement of the efficient storage and disposal of agricultural wastes and silage to regenerate nature woodlands and heather moors. For Scotland, see *Scottish Office Agriculture and Fisheries Department*.

Montreal Protocol 1987

The Montreal Protocol on Substances that Deplete the Ozone Layer set defined targets for the reduction and eventual elimination of ozone-depleting substances, which came into force on January 1, 1989. The Protocol was made pursuant to the 1985 framework Vienna Convention for the Protection of the Ozone Layer. It lists two groups of substances with ozone-depleting potential: Group I, comprising five chlorofluorocarbons (CFCs) *(qv)*, and Group II, comprising three halons *(qv)*. Parties are required to freeze and reduce production of these substances over an agreed timetable. Further substances were added to the groups by the London amendments (1990), which also revised the timetable for compliance, and again by the Copenhagen amendments (1992) and the subsequent meetings of the parties. The European Communites (EC) has obligations under the Vienna Convention and has adopted Regulation 91/594 on substances that deplete the ozone layer, which is in some respects more stringent than the Protocol itself.

mosses

Plants of the class Cryptogamia which grow under damp conditions, leading in areas of high rainfall to the formation of blanket bogs and peat.

municipal solid waste (MSW)

Refuse collected by waste collection authorities (or by contractors working for the authorities); normally composed of household waste and an element of commercial waste from shops and offices for which a charge is made.

mutagen

An agent capable of modifying genes, which are the materials of heredity.

mutagenic

Capable of causing a change in the amount or structure of deoxyribonucleic acid *(qv)* resulting in an inheritable change in the characteristics of a cell or organism.

mutation

A transmissible change in the structure of a gene or chromosome.

nanogram

One thousandth of one millionth of a gram.

National Accreditation Council for Certification Bodies (NACCB) Environment Action Panel

A panel which agrees the accreditation criteria for those who will be assessing companies and awarding certificates under the European Communities' (EC) Eco-management and Auditing Scheme (EMAS) *(qv)* and BS 7750. The NACCB released its consultation document on the criteria in October 1994.

National Association of Waste Disposal Contractors (NAWDC)

The incorporated trade association of the UK commercial waste management industry to which most of the major companies subscribe. Founded in 1968, with the principal objectives of improving operational standards overall in waste management, and representing its members' views to Government. Provides training in waste contracting, management, engineering and transportation disciplines. It also issues Codes of Practice and Conduct. Membership can only be achieved after a formal approval procedure. [*Address: 6–20 Elizabeth Street, London, SW1W 9RB; T: 0171 824 8882; F: 0171 824 8753.*]

national nature reserve (NNR)

A reserve designated under the Wildlife and Countryside Act 1981, s.35. Such reserves are of national importance and designation is by the appropriate national nature conservancy council: English Nature *(qv)*, Scottish Natural Heritage *(qv)* or the Countryside Council for Wales *(qv)*. The designated land must already be subject to a management agreement *(qv)* with the appropriate conservancy council. There are 121 NNRs in England (41,312 ha); 46 NNRs in Wales (12,798 ha) and 70 NNRs in Scotland (114, 478 ha).

All NNRs are designated sites of special scientific interest (SSSIs) *(qv)*. Some are also designated, or are candidates for designation, as special protection areas (SPAs) *(qv)* under EC Directive 79/409 on wild birds; as special areas for conservation (SACs) *(qv)* under EC Directive 92/43 on habitats; and may have international designation under the Berne *(qv)* or Ramsar Conventions *(qv)*. See also *Habitats Directive.*``

national park

An area designated for the purpose of preserving and enhancing the natural beauty of the tracts of country in England and Wales that by reason of:

1 their natural beauty, and

2 the opportunities they offer for open-air recreation, having regard both to their character and to their position in relation to centres of population, it is especially desirable that the necessary measures should be taken (National Parks and Access to the Countryside Act 1949, s.5).

The Environment Bill 1995 proposes the new purposes of 'conserving the natural beauty, wildlife and cultural heritage' of the areas, and 'promoting opportunities for the quiet enjoyment and understanding of the special qualities of those areas by the public'. British national parks are not owned by the State, and their management has in the past been a matter primarily for the local authorities for the areas (though two have been managed by special joint boards), with certain additional powers and finance. The Environment Bill 1995 proposes the setting up of special National Park Authorities, charged additionally with the responsibility of having regard to 'the economic and social well-being of local communities within the national park'. There are no national parks in Scotland.

national plan

A plan published by the Government in 1990 as its programme, under Article 3 of EC Directive 88/609 on large combustion plants, for the progressive reduction of emissions, notably SOx and NOx from existing plant. The plan seeks to implement the UK's obligation to reduce SO_2 emissions by 40% by 1998 and 60% by 2003, from a 1980 baseline, by allocating quotas to the major contributors in electricity generation and oil refineries. Both National Power and

PowerGen are within their overall target because of the improved market share of nuclear-generated electricity and independent generators, and also because of the switch from coal to gas. The requirements of the national plan are directly enforceable by Her Majesty's Inspectorate of Pollution (HMIP) *(qv)* in integrated pollution control *(qv)* under the Environmental Protection Act 1990, ss.3 and 7(2), notwithstanding that the plant may be in accordance with best available techniques/technology not entailing excessive cost (BATNEEC) and any statutory air quality standards.

National Radiological Protection Board (NRPB)
An advisory body established by the Radiological Protection Act 1970 which is responsible for research and development into radiation hazards and provision of advice and information on, and monitoring of, radiation, both man-made and natural, eg radon *(qv)*.

National Rivers Authority (NRA)
An independent regulatory agency established by the Water Act 1989 upon the privatisation of the water industry as part of the Government's overall commitment to achieve a regulatory separation of the powers of the water undertakers from the former service functions of the regional water authorities, principally the provision of water supplies and sewage disposal (see now *sewerage undertakers* and *water undertakers*). Its functions extend to control over abstraction from, and discharges to, controlled waters *(qv)*, and also to certain responsibilities in relation to water recreation, fisheries, flood defence and navigation. The Environment Bill 1995 proposes that it should be wholly absorbed into the new Environment Agency *(qv)*.

national scenic area
An area designated under the Town and Country Planning (Scotland) Act 1972, s.262C (as inserted by the Housing and Planning Act 1986) as being of outstanding scenic value and beauty in a national context. To some extent overtaken by Natural Heritage Areas designated under the Natural Heritage (Scotland) Act 1991, s.6.

National Society for Clean Air and Environmental Protection (NSCA)
A respected UK non-governmental organisation that campaigns on a broad front for improvements in air quality. [*Address: 136 North Street, Brighton, BN1 1RG; T: 01273 326313.*]

National Trust
A charitable trust, established privately in 1895 for 'the preservation for the benefit of the nation of lands and tenements, including buildings, of beauty or historic interest, and, as regards land, for the

preservation as far as possible of their natural aspect, features and animal and plant life . . .'.

The Trust (whose full name is the National Trust for Places of Historic Interest or Natural Beauty) was incorporated by the National Trust Act 1907 and is regulated by the National Trust Acts 1907, 1937, 1939, 1953 and 1971. It now owns or controls 230 historic houses, 580,000 acres of countryside, 545 miles of natural coastline, 130 important gardens and 60 villages.

It is funded primarily by endowment membership subscriptions (of over 2 million members), legacies and rents, and is independent of State aid, although it has benefited from grants from the National Heritage Memorial Fund. It has 3,000 staff, many specialist advisers and 27,500 volunteers. The Trust celebrated its centenary in 1995.

national waste strategy
See *Waste Strategy for England and Wales.*

Natura 2000
The collective name for sites which are the subject of EC Directive 92/43 on the Conservation of Natural Habitats and of Wild Fauna and Flora. This provides for the establishment of a coherent European ecological network of designated sites on land and at sea including both special areas of conservation (SACs) *(qv)* notified under the Directive and special protection areas (SPAs) *(qv)* notified under the Birds Directive 79/409. The overall aim of the Directive is the maintenance of biodiversity *(qv)* through the conservation of the species inhabiting designated sites.

Natural Environment Research Council (NERC)
The NERC has overall responsibility for the planning, support and encouragement of scientific research into the natural environment. Sponsored by the Department of Education, its brief encompasses various institutes who carry out the actual research, including the British Geological Survey, the British Antarctic Survey, the Institute of Oceanographic Sciences, the Institute of Freshwater Ecology and the Institute of Terrestrial Ecology.

natural gas
A mixture of hydrocarbons (principally methane, ethane, butane and propane) which is now tapped from wells under the North Sea, and which has replaced coal gas as an energy source for industrial and domestic purposes in the United Kingdom. Butane and propane are removed prior to distribution, and liquefied to form bottled gas.

Natural Heritage Area (Scotland)

An area designated by the Secretary of State for Scotland under the Natural Heritage (Scotland) Act 1991, s.6, on the advice of Scottish Natural Heritage, as being of outstanding value to the natural heritage of Scotland. See also *national scenic area.*

natural pollutants

Substances of a natural origin present in the Earth's atmosphere which may, when present in excess, be regarded as pollutants, such as ozone formed photochemically and electrical discharge; sodium chloride or sea salt; nitrogen dioxide formed by electrical discharge in the atmosphere; dust and gases of volcanic origin; soil and dust from dust storms; bacteria, spores and pollens; products of forest fires; and even excessive contamination by water, often associated with adverse weather conditions.

natural resource damages

A concept of US environmental law which is not precisely mirrored in domestic EC or UK law. Statutory liability for such damages forms part of the US Superfund legislation – specifically under s.9607(a)(4)(c) of the Comprehensive Environmental Response, Compensation and Liability Act of 1980 (CERCLA) *(qv)*. Potentially responsible parties are liable for damages for injury to, and destruction or loss of, natural resources arising from the release of a hazardous substance. The term natural resources includes land, fish, wildlife, biota, air, water, groundwater, drinking water supplies and similar resources belonging to or under the management or control of the US, state, local or foreign governments and Indian tribes (s.9601(16) CERCLA).

The concept relates to the costs of restoring or replacing damage that has occurred to the environment, and loss incurred as a result of the damage. This is a head of damages over and above clean-up costs – by which is meant the right of a government agency to recover its actual costs incurred in cleaning up a spillage or other polluting matter. Clean-up is only part of the total task of environmental restoration. A spillage may seriously damage a habitat and cause the destruction of wildlife. The task of restoring the natural environment, and of repopulating the depleted wildlife, goes well beyond simple clean-up. Indeed, the task may be impossible because the changes may be irreversible. There are also cases where losses have been inflicted beyond those redressable by restoration, and which may be incapable of ready calculation. The fair application of cost-internalisation principles requires that these costs should also be borne by the polluter. But the conversion of this logic into an appropriate legal reality has taken some time and has proved to be more politically difficult than may have at first appeared, and there have been few occasions yet on which natural resource damages have been recovered. The means of assessing such damages is highly controversial including economic techniques such as contingent valuation.

Nature Conservancy Council (NCC)

Until 1990 there was a single national Nature Conservancy Council for Great Britain, but the Environmental Protection Act 1990, s.128 established separate councils for England (the Nature Conservancy Council for England, generally known as English Nature *(qv)*), for Wales (the Countryside Commission for Wales) and Scotland (Scottish National Heritage *(qv)*). Each has responsibility for nature conservation in its own country, and all come together to form the Joint Nature Conservation Committee *(qv)* for national and international liaison.

nature conservation order

Common name for an order under the Wildlife and Countryside Act 1981, s.29, made to protect a site of special scientific interest designated under s.28 from potentially damaging operations, and which extends the period for negotiation of a management agreement *(qv)* to 12 months, or longer by agreement. If negotiations remain fruitless, the land may be subject to compulsory purchase.

Nitrates Directive

EC Directive 91/676 concerning the protection of waters against pollution caused by nitrates from agricultural sources. The Directive seeks to safeguard water quality firstly by the designation of nitrate vulnerable zones and, secondly, by the establishment of codes of good agricultural practice and programmes of training for farmers. Member States are also obliged to devise action plans for designated vulnerable zones and to monitor concentrations of nitrates in freshwaters. Control over nitrates is also included in the EC Directive 80/778 on drinking water.

nitrate sensitive areas

Areas designated under the Water Resources Act 1991, s.94 (and in accordance with EC Directive 91/676 and EC Regulation 2078/92) by the Ministry of Agriculture, Fisheries and Food and the Secretary of State for the Environment acting jointly (in Wales, by the Secretary of State for Wales) with a view to preventing or controlling the entry of nitrate into controlled waters from agricultural uses. Certain agricultural activities may be prohibited or restricted by a general order (subject to entitlement to compensation) or by an agreement with the landowner (which will bind subsequent owners) under s.95 of the Act. The Nitrate Sensitive Area Regulations 1994 (SI 1994 No 1729) list 22 designated areas in England, and prescribe the conditions under which the Minister may make payments of aid to farmers.

nitrates (NO_3)

Highly soluble salts of nitric acid, which occur naturally and are also a constituent part of man-made fertilisers. The run-off or leachate *(qv)* from nitrate-treated agricultural land can lead to water pollution resulting in eutrophication *(qv)* evidenced by algal blooms leading to oxygen deprivation and aquatic environment damage.

nitrification

The oxidation of ammoniacal nitrogen to nitrites and nitrates by nitrifying bacteria. See also *bacteria.*

nitrogen cycle

The circulation of nitrogen atoms brought about mainly by living things. Nitrogen is an essential ingredient of proteins required by all living organisms, and enters the living part of the biogeochemical cycle in two different ways:

1 atmospheric nitrogen is converted into nitrogen compounds which can be utilised by plants and animals through nitrogen fixation effected by certain nitrogen-fixing bacteria and certain blue-green algae;

2 the excrement of animals and dead bodies of animals and plants contain very complex nitrogen compounds which are broken down by ammonifying bacteria in the soil and converting it to ammonia.

Both ammonia and nitrates can be easily assimilated by plants.

nitrogen oxides (NOx)

Gases of a highly reactive nature comprising nitrous oxide (N_20), nitric oxide (NO) and nitrogen dioxide (NO_2). They form when nitrogen in air or combustion fuel is heated to over 650° C in the presence of oxygen or when bacteria in soil or water oxidise nitrogen-containing compounds. The principal man-made sources of NOx are industrial activity (such as power stations) and motor vehicles. Excessive emissions of NOx can give rise to respiratory problems in animals (including humans), deleteriously affect plant life and give rise to photochemical smog. Control over NOx emissions from major sources, such as oil refineries and electricity generating plant, is imposed by and under EC Directive 88/609 on large combustion plant, which specifies emission limits for NOx from new combustion plant (Annex VI), and sets ceilings and reduction targets for emissions of NOx from existing plant, requiring a 30% reduction for the United Kingdom by 1998 from its 1980 level. The United Kingdom Government has devised a national plan *(qv)* for achieving this target.

noise

A sound that is unwanted by the recipient. As a form of pollution, noise varies widely in its impact and effect. Some individuals are particularly susceptible to noise, others may hardly hear it. Noise may amount to an actionable nuisance, restrainable by injunction with or without damages. It may also be a statutory nuisance under the Environmental Protection Act 1990 (EPA 1990). Where noise arises from public works, such as new roads or aerodromes, there may be a right to statutory compensation under the Land Compensation Act 1973. Noise is a material consideration for purposes of development control and is the subject of PPG24, *Planning and Noise* (1994). The Noise and Statutory Nuisance Act 1993 addresses the particular problems of street noise, eg from car and premises' security alarms. The Government announced (March 1995) new measures to address the intractable and growing problem of neighbourhood noise.

noise abatement notice

A notice served under the Environmental Protection Act 1990 in respect of noise nuisance, in respect of which s.80A (inserted by the Noise and Statutory Nuisance Act 1993) makes special provision in cases of noise in the street from unattended vehicles, machinery or equipment.

Noise at Work Regulations 1989 (NAWR)

Regulations which control noise arising from all work activities, except sea-going ships, aircraft or hovercraft. Similar to the Control of Substances Hazardous to Health Regulations 1994 (COSHH) *(qv)*, they involve assessment of exposure and the provision and maintenance of adequate control measures with an emphasis on controlling noise at source as opposed to simply providing ear protection. The Regulations refer to three action levels relating to the degree of risk:

1 daily personal noise exposure of 85 dB(A);

2 daily personal noise exposure of 90 dB(A);

3 peak sound pressure of 200 pascales.

When any employees are likely to be exposed to noise above the first action level or the peak action level, the employer must ensure that the noise assessment is made by a competent person. The assessment must identify which of the employees are so exposed and provide information on how to comply with these Regulations, with an adequate assessment record. There is a requirement for every employer to reduce the risk of damage to the hearing of his employees to the lowest level reasonably practical. If employees are likely to be exposed to the second action level or the peak action level, then the employer must reduce the noise exposure other than by the provision of personal ear protectors.

noise control

The main options for noise control are those of:

1 reduction at source (eg reducing activity, using less noisy techniques, addition of reactive or absorptive silencers);

2 installing barriers which interrupt the direct path of sound (eg earth bunds or screens). Large barriers are required: for example, 10dB(A) reduction requires a 7 metre high barrier. The effect of barriers is reduced by wind, but this may be minimised by the use of absorptive barriers – these offer typically 3dB(A) improvement over non-absorptive barriers, but are more expensive;

3 the insulation of properties, eg sound insulation between flats, and double glazing with mechanical ventilation to permit windows to be kept shut. This method is used in relation to dwellings exposed to high noise levels from new motorways or other major roads, and from airports.

Further references: PPG 24 *Planning and Noise*; and MPG11, *Control of Noise from Mineral Workings*.

non-fossil fuel obligation (NFFO)

A special levy imposed by the Secretary of State for Trade and Industry, under the Electricity Act 1989, s.32, on public electricity suppliers, requiring them to buy an agreed amount of electric power generated by non-fossil fuel generating stations at a premium price. The scheme is funded out of a 10% levy on domestic electricity bills. Three orders have been made by the Secretary of State to give effect to the NFFO. These are the Electricity (Non-Fossil Fuel Sources) (England and Wales) Order 1990 (SI 1990 No 263); the Electricity (Non-Fossil Fuel) (England and Wales) (No 2) Order 1990 (SI 1990 No

1859); the Electricity (Non-Fossil Fuel) (England and Wales) Order 1991 (SI 1991 No 2490). In Scotland a parallel scheme, under a Scottish Renewables Order, is operated by the Secretary of State for Scotland.

non-governmental organisations (NGOs)

Interest groups formed for the purpose of promoting, monitoring or controlling a particular cause. NGOs with significant memberships and environmental objectives include Greenpeace *(qv)*, Friends of the Earth (FoE) *(qv)* and the Royal Society for the Protection of Birds (RSPB) *(qv)*. NGOs played a major part in and alongside the Rio Summit *(qv)*, and are involved in the United Kingdom Government's strategy for implementing its sustainable development *(qv)* programme.

non-point source pollution

Discharges into the environment other than from fixed points like chimneys or sewers, such as the seepage of fertilisers into a watercourse from agricultural operations, or water run-off from streets and paved areas. This type of pollution is less susceptible to controls over emissions than point-source pollution, and therefore more commonly tackled by controls over the sale, design or use of products (eg pesticides, fertilisers; motor vehicles). See also *diffuse source pollution*.

non-renewable sources

Natural resources which, once consumed, cannot be replaced. Mineral resources generally are regarded as wasting assets of this kind.

normal waters

A designation by the Secretary of State, under EC Directive 91/271 on urban waste water treatment, of waters to which effluent may be discharged from a sewage treatment works only if it has received both primary and secondary treatment, comprising biological treatment and secondary settlement of sewage effluent. Discharges will have to comply with concentration limits or percentage reductions to assess compliance with standards for biochemical oxygen demand *(qv)*, chemical oxygen demand *(qv)* and suspended solids. See also *less sensitive waters; sensitive waters*.

Notification of Installations Handling Hazardous Substances Regulations 1982 (SI 1357) (NIHHS)

Regulations which prohibit operations where there are notifiable quantities of a potentially hazardous substance on site unless the Health and Safety Executive *(qv)* have been notified in writing.

nuclear fission

The splitting of the nucleus of a heavy atom into two fragments of approximately equal mass, usually by neutron bombardment.

nuclear reactor

An assembly in which a fission train reaction is caused and takes place in a controlled manner. The essential components of a reactor are a fissile material for fuel, a moderator, a coolant and a neutron-absorbent material to act as a control for the neutron flux in the reactor core.

nuclear waste

See *radioactive waste.*

nucleus

The central core of the atom, composed of protons and neutrons.

nuclide

The nucleus of an isotope.

nuisance

An interference with someone's use or enjoyment of their land, or the amenities of a locality or community. The law recognises three types:

1 a private nuisance at common law;

2 a public nuisance at common law; and

3 a statutory nuisance under the Environmental Protection Act 1990 (EPA 1990).

A private nuisance is an unlawful interference with a person's use or enjoyment of land, or some right over, or in connection with it. It is concerned with infringement of private property rights, and this tends to limit its applicability as a means of redress for environmental harm. To succeed in private nuisance, a plaintiff must have an interest in the land affected by the nuisance. Thus private nuisance will not afford a remedy to members of the occupier's family, or to a licensee without possession such as a guest or lodger. Some recent case law has affirmed the *Malone v Laskey* principle. See for example: *Hunter v Canary Wharf* (*Independent,* December 20, 1994) contra *Khorasandijian v Bush (The Times,* February 19, 1993).

Public nuisance is both a tort *(qv)* (ie civil wrong) and a crime. It has been defined as a nuisance 'which materially affects the reasonable comfort and convenience of life of a class of Her Majesty's subjects' (*Attorney General v PYA Quarries Limited* [1957] 2 QB 169). Unlike private nuisance, a proprietary interest in land is not required but a plaintiff will have to show that he has suffered special (ie substantial, direct and not consequential) damage over and above that suffered by the relevant class (ie group of people) as a whole. In environmental claims both private and public nuisance actions are often pursued for the same wrong, eg *Halsey v Esso Petroleum* [1961] 1 WLR683, where damages and injunctive relief were successfully sought for noise nuisance, smells and damage by acid smuts from the defendants' oil depot.

A statutory nuisance *(qv)* is a statutory form of common law nuisance, specified in Part III of the Environmental Protection Act 1990 (EPA 1990). In summary this provides a simplified procedure for a local authority to bring abatement proceedings for specific types of nuisance (s.79(1)) or for an aggrieved individual to complain direct to a Magistrates' Court seeking an order abating the nuisance (s.82). An important proviso in proceedings for statutory nuisance is that the alleged nuisance must also constitute a nuisance at common law – which may sometimes impose some of the legal problems inherent in establishing common law nuisance.

See also *Cambridge Water Company case; Rylands v Fletcher; statutory nuisance.*

nutrient

See *eutrophication.*

nutrient stripping

A tertiary treatment of waste waters, either to reduce the rate of eutrophication of receiving waters, or to permit the reuse of water for domestic purposes. Methods range from chemical coagulation to advanced treatment processes, such as those developed for the desalination of sea water and brackish waters.

occupational exposure standard (OES)

A standard prescribed by the Health and Safety Commission *(qv)*, under the Health and Safety at Work Etc. Act 1974, for the control of exposure to a particular substance. The OES is not to be exceeded; and if it is exceeded, action must be taken to remedy the situation as soon as is reasonably practicable.

occupier

A person who has physical use of land without necessarily owning either a freehold or leasehold interest in it. The exact meaning of the term varies depending on the legislative context. Liability to remedy environmental harm is normally imposed on the person who caused the pollution in the first place, but in relation to ongoing damage arising from land it is common also to attach liability to the person now in control of the land, as owner or occupier.

offensive industry

Any business which, by reason of the process involved or the method of manufacture, or the nature of the raw materials or goods used, produced or stored, is likely, as a result of the inadequacy of the available means of control, to cause effluvia, fumes and odours offensive to persons on adjacent land.

Office of Gas Services (OFGAS)

The economic regulator of the privatised gas industry, created in 1986 and headed by the Director General of Gas Supply. It has two main aims: to protect the interests of the gas consumer and to facilitate the introduction of competition in the domestic gas market planned to take place between 1996–98.

Office of the Director General of Electricity Supply (OFFER)

The economic regulator of the privatised electricity industry, under the Electricity Act 1989. The Director General shares with the Secretary of State the duty to take into account, in exercising his regulatory functions, the effect on the physical environment of activities connected with the general transmission or supply of electricity.

Office of Water Services (OFWAT)

The office of the Director General of Water Services, appointed under the Water Industry Act 1991, s.1, who is responsible for monitoring the performance of the water undertakers *(qv)* and sewerage undertakers *(qv)* to secure economy and efficiency and to protect the interests of customers with regard to charging, prices and standards of service. The Director General may investigate complaints from customers about water and sewerage companies and adjudicate in some disputes but he mainly takes the appropriate initiative himself. His principal function is in setting maximum charges which the undertakers may levy for water services, and he has been required to have regard to their obligations under national and EC environmental law.

In the 1994 periodic review, OFWAT insisted that price increases to consumers over the next 10 years should be limited to what was necessary to meet statutory environmental obligations, and that companies would be expected to carry out new obligations economically at a modest return on new investment, and to absorb a substantial part of the cost by increasing operating efficiency and by managing a lower return on their existing capital base. The major component in future environmental costs is the implementation of the Urban Waste Water Treatment Directive *(qv)*, which is estimated to cost around £6 billion at 1993–94 prices, plus a further £0·8 bn for discharges to bathing waters. Further reference: OFWAT, *Future Charges for Water and Sewerage Charges* (1994).

oil pollution

Environmental damage caused by escapes of oil, including crude oil, fuel oil, sludge, oil refuse and refined products. A dramatic example was the *Torrey Canyon* disaster in 1967 which caused contamination to miles of coastline and wildlife devastation (see also *Exxon Valdez* and *Amoco Cadiz*). As a consequence, the Merchant Shipping (Oil Pollution) Act 1971 was passed to give effect to international obligations on civil liability for oil pollution damage. Under the Merchant Shipping (Prevention of Oil Pollution) Regulations 1983 (as amended) ships within UK territorial waters must not discharge oil or any oil mix into the sea.

opencast mining

The winning of materials through surface excavation. Overlying soils and rocks (overburden) are removed first to provide access to ores, coal and other minerals. Opencasting is governed by the Opencast Mining Act 1958, and, since 1981, under planning legislation. Guidance is contained in MPG3, *Coal Mining and Colliery Spoil Disposal* (1994).

organic

Containing carbon compounds.

Organisation for Economic Co-operation and Development (OECD)

An organisation of developed countries which was established in Paris in 1960, and which seeks to achieve the highest sustainable economic growth and employment together with a rising standard of living in Member countries, whilst maintaining financial stability, and thus contributing to the development of the world economy; and to maximise the expansion of world trade on a multilateral, non-discriminatory basis in accordance with international obligations. The OECD consists of 24 member states including those of Europe, Canada, Japan and the US. The council is the executive arm whose decisions (which are binding where there is consensus) may lead to the introduction of new national laws. The OECD played an important role in development of the polluter pays principle *(qv)*, and has published studies of the use of economic instruments *(qv)* for environmental protection.

organochlorine

A compound in which carbon atoms are linked directly with chlorine atoms. Includes many pesticides, such as pentachlorophenol (PCP) and dieldrin (*qv*).

orimulsion

A cheap bitumen-based fuel imported from Venezuela, which is used as a substitute for coal in electricity generating plants. Its combustion produces fine particulates *(qv)* and fumes containing nickel and vanadium, which require special gas scrubbing equipment.

osmosis

A process which occurs when water or other liquids flow through a semi-permeable membrane which is a solid sheet or film which permits the passage of the liquid but not the substances dissolved in it. The general tendency is for the fluid to flow from a weaker to a stronger solution.

owner

The person who owns, holds or has the rightful claim or title to an object. The concept is of significance in environmental law because ownership of land tends to carry with it the necessary control to remedy environmental harm, so the owner is an appropriate person against whom enforcement action can be taken, even although the harm may have been caused originally by a previous owner or occupier, or some other third party. In some statutory contexts, 'owner' is given a special meaning so as to include the owner of the freehold, but to allow liability to shift to someone who holds under a long lease, and who has a capital interest in the land.

oxidation pond

A shallow lagoon or basin within which waste water is purified, through sedimentation, and both aerobic and anaerobic biochemical activity over a period of time, providing the climate is favourable.

ozone

Ozone is a form of oxygen, each molecule of which consists of three oxygen atoms. The presence of ozone in the stratosphere is important because it absorbs solar ultraviolet radiation which is potentially damaging to the environment and to human health. In the lower atmosphere it acts as a greenhouse gas and at ground level it is a pollutant, harmful to plants and a contributor to photochemical smog. Ozone is a powerful oxidant and is used to remove low-level chemical and microbiological impurities in water.

ozone-depleting substances

Chemical substances of natural and anthropogenic origin which are thought to have the potential to modify the chemical and physical properties of the ozone layer. The substances with the most damaging ozone-depleting potential include CFCs, halons, carbon tetrachloride and trichloroethane.

ozone layer

A layer of the atmosphere surrounding the Earth at a distance of between 10km and 50km above its surface. It is not a static layer since ozone is constantly being created by the action of ultraviolet radiation (a component of sunlight) on oxygen molecules. At the same time it is being destroyed by a variety of chemical reactions, many of which derive from gases emitted by both natural and human sources on Earth. There is, as a result, a dynamic process of production and destruction of ozone in the stratosphere. Recently the rate of destruction has exceeded the rate of production leading to a thinning in the ozone layer, with potentially adverse consequences for human health, organisms and ecosystems.

packaging

A case or other container used to package goods; any box, packet or other article in which one or more product is enclosed; the total presentation of a product for sale including its design and wrapping. A draft EC Directive (93/416) on packaging and packaging waste proposes rules to encourage the recovery and recycling of packaging waste. See also *Duales System Deutschland* and *producer responsibility*. Further reference: House of Lords Select Committee on the European Communities, *Packaging and Packaging Waste* (HL Paper 118; 1992–93).

paragraph

A term used to denote a sub-division of legislation, such as a statute (it is given to sections of a Schedule, and to sub-divisions of a sub-section within the main body of the Act); a regulation (where each numbered part is itself called a regulation, but a sub-division (eg reg. 5(6)) is a paragraph; an order (where each numbered section is termed an article, but sub-divisions are called paragraphs. See also *article*.

parameter

Either an arbitrary constant or a characteristic constant of a mathematical expression or statistical population, such as the variants or the mean of a population, or a general description given to a variable.

Paris Convention

The Convention for the Prevention of Pollution from Land-based Sources, concluded in Paris in 1974, which addressed the problems of persistence, toxicity and bio-accumulation of certain pollutants entering the marine environment. It was ratified by the European Community in 1975 and extended to airborne pollution in 1987. Its administration is the responsibility of the Paris Commission.

particulates

Very small solid or liquid particles suspended in the atmosphere. Some solid particles can transport pollutants other than their own due to their irregular shape. For example, smoke particles and sulphur dioxide have greater effects on health when emitted in combination rather than when separate. A full study of diesel particulates is contained in the Royal Commission on Environmental Pollution's *15th Report: Emissions from Heavy Duty Diesel Vehicles* Cm 1631 (1991).

pathogen

An organism capable of causing a disease, typically a micro-organism. Examples are staphylococci, streptococci, coliforms and salmonellae. These types of pathogens can be found in or around waste disposal sites. They include viruses, such as enteroviruses, polioviruses and hepatitis A virus, all of which can be spread by air or through dust.

penalties

Punishments imposed for breach of a law; they may include pecuniary sanctions (for example a fine) or other disabilities or disadvantages fixed by law for some offence, for example imprisonment or disqualification under the Company Directors Disqualification Act 1986.

pentachlorophenol

A bactericide used as an industrial wood preservative and in the treatment of leather and other textiles. Often referred to as PCP. It tends to be absorbed into soil and can be degraded by photolysis in water, which is a reaction involving light. It is very slow to biodegrade, and hence is persistent in soils. It is a UK red list (*qv*) substance.

perceived risk

A subjective appreciation by individuals which will more often than not bear little relation to the statistical probability of damage or injury.

Risks that seemingly are under voluntary control are considered as less potentially hazardous than those over which a person seems to have no control. The chance of being harmed through playing sports or in a road accident is, therefore, thought to be far less than the possibility of damage or injury resulting from a nuclear power plant although the risk of damage or injury resulting from playing sports or a road accident is many, many times greater than the latter.

percentile

A parameter set for compliance of samples taken to test for compliance with regulatory limits, for example in relation to water quality, and which requires that some percentage of samples (eg 95-percentile) should satisfy the maximum admissible concentrations prescribed.

percolating filter

An artificial bed over which sewage is distributed and through which it percolates to underdrains, thus giving an opportunity for the formation of biological slimes which bring about oxidation and clarification of the sewage. Sometimes referred to as a trickling filter or a bacteria bed.

permitted development rights

see *General Development Order.*

person aggrieved

See *locus standi.*

personal protective equipment (PPE)

All equipment, including clothing, affording protection against all hazards in the workplace which is intended to be worn or held by a person at work and which protects him against one or more risks to his health and safety, and any additional accessory designed to meet that objective as prescribed in the Health and Safety (Personal Protective Equipment at Work) Regulations 1992 (1992/2966). The

Health and Safety at Work Etc. Act 1974 provides that an employer's primary duty is to ensure, so far as is reasonably practicable, the health, safety and welfare whilst at work of all employees. Personal protective equipment should be relied on by employers only if it is not reasonably practicable to alter the normal work practices themselves in order to avoid placing an employee in a situation where he or she will need this type of equipment. Any equipment so provided must be suitable for the purpose for which it is intended. See *Control of Substances Hazardous to Health Regulations 1994.*

pesticides

Substances or products including insecticides and fungicides which are used to control pests. There are three main types:

1 Chlorinated hydrocarbons (eg dichloro diphenyl trichloroethane (DDT) *(qv)*) which are used to control mosquitoes and houseflies. They are not particularly toxic to mammals but tend to concentrate in the fatty tissues especially around the vital organs. The biodegradability of chlorinated hydrocarbons is 75–100% disappearance in four years.

2 Organophosphates, which are short-lived. They can be used in agriculture, gardening and public health. They can be microbially degraded. The biodegradability of organophosphates is 75–100% disappearance in one week.

3 Pyrethins, which were originally based on natural sources from pyrethrum flower heads but are now being artificially synthesized in very large amounts. They are relatively safe compared to the two previous classifications.

See also *mercury.*

petrochemicals

Intermediate chemicals derived from petroleum, hydrocarbon liquids or natural gas. They are made up of mainly carbon and hydrogen but also nitrogen, sulphur and oxygen and minor proportions of potassium, vanadium and nickel. The term includes a wide variety of organic compounds, for example:

1 aliphatic hydrocarbons;

2 aromatic hydrocarbons;

3 resins and asphaltenes.

pH

A measure of the acidity or alkalinity of a solution, numerically equal to 7 for neutral solutions, increasing with increasing alkalinity and decreasing with increasing acidity. On this scale a value of 0·0 is highly acid, and a value of 14·0 is highly alkali. Originally stood for the words 'potential of hydrogen'.

phenol

A member of the class of aromatic organic compounds in which one or more hydroxy (OH) groups are attached directly to a benzene ring. Examples are phenol itself (also called carbolic acid, benzophenol or hydroxybenzene), the cresols, xylenols, resorcinol and naphthols. Though technically alcohols, their properties are quite different due to the benzene ring. Phenols are toxic by ingestion, inhalation and skin absorption, and can also be a strong irritant to tissue. They are capable of degradation by natural micro-organisms in soils, but can be a particular problem at low levels in water supplies, as they can migrate through some plastics and form strongly tasting chlorophenols in drinking water supplies.

phosphates

Phosphates are essential for plant metabolism. They are applied as fertiliser to rectify phosphorus deficiencies in soil. Over-application of phosphates can result in pollution of rivers causing eutrophication (*qv*) (which results in algae growth, causing plants and animals to die as well as the possible contamination of groundwater).

photochemical reaction

A chemical reaction which is initiated by light of a specific wavelength. In an environmental context an example is the potential action of ultraviolet light on CFCs which may bring about the detrimental degradation of the ozone layer. Photochemical reactions initiate the process of photosynthesis *(qv)* in which plants convert carbon dioxide into sugars which are incorporated into cell materials (with the release of oxygen).

Another possible use of an advantageous photochemical reaction is in the biodegradation of plastics. In 1992 the plastics industry alone produced nearly 70 billion pounds of plastics, 40% of which was put into landfills. Due to the persistent nature of many of these synthetic polymers, recent steps have been taken to produce photo-biodegradable starch-linked polymers whose structure alters under exposure to ultraviolet light provided by sunlight, making them susceptible to attack by micro-organisms and hopefully resulting in consequent degradation.

photosynthesis

The process by which plants capture the natural energy of the sun by converting carbon dioxide into carbohydrates. It takes place in the chloroplast of green plants. The oxygen which is released helps to maintain the oxygen content of the atmosphere and the carbohydrate produced is a source of energy for plants and animals.

phytotoxic

Poisonous to plants.

plankton

Unicellular eukaryotes which are fed by engulfing particulate matters.

planning advice note (PAN)

The Scottish equivalent of planning policy guidance *(qv)*.

planning control

See *development control.*

planning gain

The expression that remains in common use (despite the Government's disapproval of it in DOE Circular 16/91, *Planning and Compensation Act 1991: Planning Obligations* paras B2 and B3) to describe the outcome of negotiation between a local planning authority *(qv)* and a developer in which some asset or right is transferred to the authority or to the local community in return for the grant of planning permission *(qv)*. Examples include a contribution towards the cost of widening or improving local roads to accommodate traffic generated by the proposed development, and perhaps to make up shortfalls in the existing infrastructure.

Government guidance alludes also to the prospect of planning gain being used for environmental protection purposes, such as maintaining and protecting from development an area of habitat importance; and planning gain is likely to play an important role in negotiations for redevelopment of sites which are, or are near, land which is contaminated. In a series of recent cases the courts have taken a relaxed view of planning gain, and shown a willingness to uphold such deals provided their benefits are still capable of constituting a 'material consideration' for the purposes of development control *(qv)*: see eg *R v Plymouth City Council and others, ex parte Plymouth and South Devon Co-operative Society Limited* [1993] 2 PLR 75; *Tesco Stores Limited v Secretary of State for the Environment and others* [1994] JPL 919; *Crest Homes Ltd v Secretary of State for the*

Environment (Court of Appeal; October 1994). If planning gain cannot lawfully be included as a condition in a planning permission it is likely to be effected by way of a planning obligation entered into under the Town and Country Planning Act 1990, s.106. A planning obligation binds the land and is registrable as a local land charge.

planning permission

Permission granted by a local planning authority *(qv)* (or, on appeal or call-in, by the Secretary of State) to carry out development *(qv)* on land. Planning permission is also granted by the General Development Order *(qv)* for a range of types of development. Special additional rules and procedures apply in the case of listed buildings *(qv)* and buildings within conservation areas *(qv)*, and also to the storage of hazardous substances *(qv)*. It is unlawful to undertake development without planning permission, or to fail to comply with any condition imposed on the grant of planning permission, entitling a local planning authority to issue an enforcement notice *(qv)* specifying the breach of planning control alleged by them, and requiring it to be rectified in a specified way and by a specified date, on pain of criminal penalties.

Planning permission may be full or, in the case of applications for building work, in outline. Outline planning permission is often sought by applicants wishing to establish the principle of development without going to the expense of preparing detailed plans; usually matters such as siting, design, access, landscaping and external appearance are reserved for further approval when detailed plans have been prepared. Application for approval of these reserved matters needs to be made within three years of the grant of the outline planning permission (unless the local planning authority substitutes a shorter or longer period). Full planning permission has to be implemented within five years of its grant unless a shorter or longer period is substituted.

Planning control plays a particularly important part in environmental protection in the United Kingdom, for the following reasons:

1 There is no entitlement to compensation for refusal of planning permission, which makes planning control particularly strong as a tool for environmental protection, especially in relation to green belts and habitats for flora and fauna.

2 In determining planning applications, local planning

authorities are not bound by their development plans, but are required to have regard to all material considerations; these include general and specific environmental considerations. Whilst the plan must be followed unless these material considerations indicate otherwise (Town and Country Planning Act 1990, s.54A), environmental protection is capable of being a theme both of development plans and of development control. Expression is given to these interests and the overlap between planning and other environmental protection laws in PPG23, *Planning and Pollution Control* (1994).

3 Planning is the mechanism through which environmental impact assessment is for the most part implemented in the United Kingdom, because its comprehensive character means that there are few projects subject to the EC Directive 85/337 on environmental assessment which do not also require planning permission (special regulations have been made for cases such as harbour works, afforestation, highways and salmon fishing, all of which, for different reasons, fall outside normal planning controls).

planning policy guidance (PPG)

Planning policy guidance on the Government's policies is contained in published notes and guidance on many different aspects of planning. The first PPGs appeared in 1988 and the intention was for them to provide a comprehensive statement of Government policy on planning, which previously was issued in circulars (now confined primarily to procedural advice). The notes have particular importance to aspects of environmental protection, and the role of the planning system in this respect, especially PPG9, PPG22, PPG23 and PPG24.

The full list of current guidance is as follows: PPG1, *General Policy and Principles* (1992); PPG2 (revised), *Green Belts* (1995); PPG3, *Housing* (1992); PPG4, *Industrial and Commercial Development and Small Firms* (1992); PPG5, *Simplified Planning Zones* (1992); PPG6, *Town Centres and Retail Developments* (1993); PPG7, *The Countryside and the Rural Economy* (1992); PPG8, *Telecommunications* (1992); PPG9, *Nature Conservation* (1994); PPG10, *Strategic Guidance for the West Midlands* (1988); PPG11, *Strategic Guidance for Merseyside* (1988); PPG12, *Development Plans and Regional Planning Guidance* (1992); PPG13, *Transport* (1994); PPG14, *Development on Unstable Land* (1990); PPG15, *Planning and the Historic Environment* (1994); PPG16, *Archaeology and Planning* (1990); PPG17, *Sport and Recreation* (1991); PPG18, *Enforcing Planning Control* (1992); PPG19, *Outdoor Advertisement Control* (1992); PPG20, *Coastal Planning* (1992); PPG21, *Tourism* (1992); PPG22,

Renewable Energy (1993); PPG23, *Planning and Pollution Control* (1994) and PPG24, *Planning and Noise* (1994).

plastics recycling

The difficulty with the recycling or recovery of plastics is that the greater volume of plastics arise from municipal solid waste and become contaminated with other materials and therefore uneconomic to separate. The greater volume of plastics, therefore, are presently destined for landfill. The proportion of paper to plastics and consequently the calorific value in municipal waste is increasing, albeit slowly. Thus although there are certain specialised plastic recovery units, such as for plastic containers, the general anticipation is that the contribution that plastics can best make towards recycling and recovery is through their calorific value. The average calorific value as received at an municipal solid waste incinerator for paper is 14,600 Kj/Kg which is 6,250 Btu/lb as opposed to plastic which is 37,000 Kj/Kg which is 16,100 Btu/lb.

playing field, level

A mythical recreational facility upon which businesses and nations compete without handicap or disadvantage.

pm 10

Particles of less than 10 microns in size, capable of penetrating deep into lungs. Currently believed to be one of the constituents of vehicle exhausts, particularly diesels, most difficult to eradicate or filter.

point source pollution

Pollution from a discrete source, such as a septic tank, a sewer, a discharge pipe, a landfill, a factory or waste water treatment works discharging to a watercourse; stack emission from an industrial process; or spillage from an underground storage tank leaking into groundwater. See also *diffuse source pollution*; *non-point source pollution*.

polishing processes

Additional treatments which are often designed to produce effluent better than the Royal Commission on Environmental Pollution's *(qv)* standard of suspended solids at 30 mg/l and biological/biochemical oxygen demand (BOD) *(qv)* at 20 mg/l in water reclamation processes.

pollutant

A substance, whether solid, liquid or gas, which when introduced into the environment adversely affects all or any part of the ecosystem to an unacceptable extent. Pollutants can interfere with chemical balances, growth rates of species, food chains, property values, health, amenity and comfort. Pollutants, eg sulphur dioxide, can be assimilated into biological cycles and are usually biodegradable or they are synthetic (eg chlorinated hydrocarbons may be not biodegradable and can accumulate in the ecosystem).

polluter pays principle

The principle adopted by Organisation for Economic Co–operation and Development (OECD) *(qv)* countries in 1972, incorporated in the EC's first Environmental Action Programme (*qv*), and now contained in the Treaty of Rome (Article 130r(2)), that polluters should bear the full cost of prevention and minimisation of pollution, and of remedying environmental damage, and that this cost should be reflected in the cost of goods and services which cause pollution in their production, consumption or disposal.

polluting emissions register

A proposal by the European Commission to introduce a register to enable industry to report on the emission levels of various substances believed to be harmful to health and to the environment. The development of a common emission reporting standard throughout the Communities may well take three to four years and almost certainly longer to achieve the overall necessary consistency in all EC countries. It is the intention of the Government that the new Environment Agency should follow up on the work of the US Toxic Substances Inventory (TSI) and by 2000 ensure that all major UK businesses of over 200 people report in writing their annual pollution emissions.

pollution

A broad term, generally taken to mean harmful or unwanted effects. Whether a particular discharge into the environment causes pollution depends on the circumstances in which it occurs, the capacity of the receiving environment to dilute and disperse, the nature and the quality of the substance, assessment of risks, public attitudes and value judgements. Further reference: Royal Commission on Environmental Pollution, *10th Report; Tackling Pollution – Experience and Prospects* (Cmnd 9149; 1984).

Another definition of pollution, which has achieved common currency and is well worth consideration, is the introduction by man into the environment of substances or energy liable to cause hazards to human health, harm to living resources and ecological systems, damage to structure or amenity, or interference with legitimate uses of the environment. See *A Perspective of Environmental Pollution,* MW Holdgate, Cambridge University Press, 1979.

There has always been controversy as to whether contamination can be distinguished from pollution. Contamination generally describes situations where the presence of substances is believed or positively asserted to be harmless, while pollution implies actual damage. The Royal Commission preferred to use 'contamination' to imply the presence of alien substances or alien energy in the environment without passing judgement on whether they cause or are liable to cause damage. Contamination is thus a necessary, but not a sufficient, condition for pollution.

polychlorinated biphenyl (PCB)

An organochlorine compound that is very toxic and has a high potential for bio-accumulation. Because they are stable chemically, PCBs have been widely used as transformer coolants, wire and cable coatings and insulating materials. PCBs have been used in industrial sealing compounds, adhesives, plastics and rubber, insecticides, paints and varnishes and in the surface coating industry. They were first produced in the US in 1929, and world production today totals approximately 1·2 million tonnes. Only 4% of the PCBs in the environment have been degraded, 31% exist as residues and 65% are landlocked in use with existing chemical plants, storage and landfill. Concern about PCBs in the environment is renewed by the fact that whilst the compounds are biologically stable, they are lipophilic (ie soluble in fat). PCBs can be carcinogenic and mutagenic depending upon their degree of chlorination and structure and can be accumulated in the fatty tissues of certain fish, marine mammals and birds. These are passed on to offspring since they are not easily broken down by the species' metabolism. PCBs have been banned from production since 1979, but are still contained in many electrical components. These components require specialist disposal, by incineration, when they eventually come to the end of their useful life.

The Control of Pollution (Supply and Use of Injurious Substances) Regulations 1986 (SI 1986 No 902, amended by the Environmental Protection (Controls on Injurious Substances) Regulations 1992, SI 1992 No 31) have prohibited new use of PCBs in the United Kingdom since 1986. At the Third International Ministerial Conference on the Protection of the North Sea, 1990, Belgium, Denmark, France, Germany, The Netherlands, Norway, Sweden and the United Kingdom agreed to phase out all remaining identifiable uses of PCB by 1995 with a complete phase-out by December 31, 1999. Further reference: Department of the Environment *Waste Management Paper No 6, Polychlorinated Biphenyls.*

polychlorinated dibenzo-p-dioxins and furans (PCDDs and PCDFs)
See *dioxins; furans.*

polychlorinated hydrocarbons (PCH)
In general the more chlorinated the hydrocarbon the greater is its resistance to degradation. Highly chlorinated hydrocarbons are attacked by anaerobic micro-organisms (no oxygen) as opposed to aerobic organisms and a by-product of this is vinyl chloride which is highly toxic. Tetrachloroethylene is a banned substance in the Montreal Protocol *(qv)* as it is a potential ozone degradative agent.

polycyclic aromatic hydrocarbons (PAH)
Neutral, non-polar organic compounds of two or more fused benzene rings and may originate from any petrochemical source such as from a North Sea oil complex to a disused coke works site. Concern about PAH pollution is founded upon their toxicological effects since they can be carcinogenic, teratogenic and mutagenic. The main problems arise with larger PAH molecules but recent research has shown that enhanced degradation of these may be achieved through oxidation and microbial activity.

polyethylene terephthalate (PET)
A transparent, plastic-like substance increasingly used for the manufacture of beverage containers.

polymerisation
A process used for treatment of those liquid wastes with a high stabilisation solids content. Stabilisation is a process in which a waste is converted to a more chemically stable form. Solidification is a process in which materials such as fly ash or Portland cement are mixed with waste to produce a slurry which sets to a cement-like solid. Polymerisation is a combination of these two processes and where the resultant product is landfilled. Questions have been raised, however, concerning the stability of the final product and the possibility of leaching at a future date. The House of Commons Environment Committee in 1989, in its report on *Toxic Waste,* considered the availability of these processes essential in waste management, offering significantly enhanced environmental protection as compared with direct landfill of potentially toxic materials provided that there is adequate technical back-up available to the operator.

polyvinyl chloride (PVC)

A substance used in rigid materials (eg for construction), flooring, electrical coverings, clothing, furnishings, packaging, toys and luggage, and which causes problems for waste disposal. The presence of chlorine in PVC alters the properties of polyethylene to make the product smooth, shiny and impermeable to gases, oils, fats and flavour. PVC can be softened and shaped by heating and then hardened by cooling, but PVC powder is a potential health hazard and is reported to be the cause of pneumoconiosis. The chlorine contained in PVC is non-combustible and upon incineration hydrochloric acid is produced.

potable water

Water suitable for use as drinking water *(qv)*.

precautionary principle

The principle that even when the (exact) effect, or even whether there is any effect, of a potentially harmful emission or discharge into the environment is not known, a presumption exists against its release. The principle is a central plank of EC environmental policy and is now incorporated into the Treaty of Rome (Article130r(2)), although it does not have direct effect *(qv)* in United Kingdom law (see *electromagnetic fields*).

precipitation

In chemistry, this is the separation and deposition of a substance in a solid form from solution in a liquid, through the action of a chemical reagent or of electricity, heat, etc. In the context of physics and meteorology, it is the condensation and deposition of moisture from the state of vapour, through cooling, particularly in the formation of dew, rain, snow, etc.

predict and provide

An approach to the planning of infrastructure or other facilities based on meeting projected demand; often associated with demand projections themselves based on simplistic extrapolation from past trends. Heavily criticised by opponents of road schemes, power stations, dams and the like for failing to take into account, or for concealing a decision to reject, the opportunity which major investment decisions provide to alter, as well as merely to confirm, past trends in and the existing pattern of society.

prescribed processes and prescribed substances

Processes and substances defined in detailed regulations made under the Environmental Protection Act 1990, s.2 for the purposes of integrated pollution control *(qv)* and local authority air pollution control. The Environmental Protection (Prescribed Processes and Substances) Regulations 1991 (SI 1991 No 472) (as amended) provide that no prescribed processes may be carried out pursuant to which prescribed substances are released into any environmental medium, without authorisation. Schedule 1 to the 1991 Regulations specifies in detail the prescribed processes and Schedules 4–6 specify the prescribed substances. See also *air pollution control.*

prior informed consent (PIC)
Prior informed consent (PIC) is required to be given by non-member states before any export may be made to them, by a Member State, of those chemicals which are banned or severely restricted within the EC. The principle of PIC has been incorporated into international agreements such as the Basle Convention *(qv)* concerned with waste materials and by the United Nations Environment Programme and Food and Agriculture Organisation. Banned and severely restricted chemicals include mercury, organochlorine, pesticides, asbestos and polychlorinated biphenyl. The forerunner of PIC was the principle of prior informed choice which obliged Member States to give notice to intended countries of destination when exporting banned or severely restricted substances. Under the principle of prior informed choice, however, as opposed to PIC, a shipment could go ahead if the importing country failed to respond to a notification within a prescribed period.

priority natural habitat types
Natural habitats which are in danger of disappearance, within the EC and for which the EC has a particular responsibility. They are referred to in Article 1 of the EC Habitats Directive (*qv*) now implemented by the Conservation (Natural Habitats, &c.) Regulations 1994 (SI 1994 No 2716). Priority natural habitat types are indicated with an asterisk in Annex I of the EC Habitats Directive eg continental salt meadows; Mediterranean temporary ponds; Caledonian forest; bog woodlands.

private Bill procedure
A parliamentary process under which any citizen can secure the enactment of a special measure, subject to proving that the end sought can be obtained in no other way. Widely used for railway works over the past two centuries, in the absence of any general enabling legislation for the construction or alteration of railways and associated land acquisition and rights of access; and also for works, such as marinas and barrages, affecting public rights of navigation. Authorisation of major projects by private Bill is exempted from the

requirements of EC Directive 85/337 on environmental impact assessment. Following concern about the appropriateness of parliamentary processes compared with public local inquiries, and the potential for private Bill procedure to override statutory planning and environmental controls, Parliament enacted the Transport and Works Act 1992, which established a new procedure, involving where appropriate environmental assessment and a public local inquiry, for almost all projects formerly requiring authorisation by private Bill. See also *hybrid Bill.* Further reference: *Report of the Joint Committee on Private Bill Procedure* Session 1987–88; HL Paper 97; *Private Bills and New Procedures – A Consultation Document,* Cm 1110 (HMSO 1990).

process standard

An environmental protection standard relating to the way in which a process should be conducted, and hence not directly concerned with the amount of pollutant which a factory may emit but more that the processes used should limit emission as much as is practicably possible, such as requirements relating to purification or filtration systems, a requirement for a particular pre-treatment for effluent, and specifications for the height of a chimney stack. See also *emission standard; product standard.*

producer responsibility

A concept in relation to waste, proposed under the Environment Bill 1995, which will enable the Secretary of State to make regulations imposing duties on manufacturers and others for the purpose of promoting an increase in recovery, recycling or re-use of products and materials.

product standard

A standard which prescribes aspects of the physical or chemical composition of products which have potential for causing environmental damage (such as detergents or cosmetics), or the handling, presentation and packaging of products, particularly those which are toxic. The requirements as to packaging of pesticides is one such example (Directive 78/631 which is implemented in the UK by the Classification, Packaging and Labelling of Dangerous Substances (Amendment) Regulations 1986 (SI 1986 No 192)). Regulation of the sulphur and lead contents of fuels and prohibitions on the presence of certain substances in products are other examples of such standards. Product standards can also require construction of a product in a certain way so as to limit emissions when in use. Examples of this type of product standard include the legal requirement for cars to have catalytic converters *(qv)* (Directive 91/441 which has been implemented in the UK by the Motor Vehicle

(Type Approval) Regulations). See also *emission standard; process standard.*

prohibition notice

An administrative order that may be served by a regulatory agency requiring the cessation of a specified activity. Examples in environmental protection include:

1 notices served by the National Rivers Authority under the Water Resources Act 1991, s.86 to prevent or control a discharge of effluent; and

2 notices served by enforcing authorities (Her Majesty's Inspectorate of Pollution or a local authority) under the Environmental Protection Act 1990, s.17 upon a person carrying out a prescribed process (*qv*) without, or in breach of, the necessary authorisation under Part I (integrated pollution control (IPC) (*qv*)) of that Act. The enforcing authority may serve a prohibition notice where a condition in an IPC authorisation has been contravened and in circumstances where the process is found to involve 'an imminent risk of serious pollution to the environment' (EPA 1990, s.14). Unlike enforcement notice under EPA 1990, s.13, prohibition notices can aim at curtailing a legal activity if new evidence comes to light that the continuation of the process is likely to cause serious pollution.

Appeal against prohibition notices is to the Secretary of State for the Environment (EPA 1990, s.15). Failure to comply with a prohibition notice is a criminal offence under EPA 1990, s.23(1)(c).

proximity principle

The principle that potential environmental damage should be contained and rectified as far as possible as close to the source of pollution. The principle seeks to avoid further environmental damage as a result of the remediation or existing environmental problems. An example is that of the transportation of waste. Waste should not be transported over long distances for disposal because of the attendant risks of accidental spillages and emissions on long journeys for waste transporting vehicles let alone the disbursement of energy (see *Commission v Belgium* – Case 2/90). A second justification for the principle in the case of waste is that it encourages communities to take more responsibility for the waste which they produce (see further PPG23, *Planning and Pollution Control* (1994); and Department of the Environment, *A Waste Strategy for England and Wales* (Consultation draft; 1995).

public health

The generally accepted justification for early environmental intervention, for example, under the Public Health Act 1936. It was the basic criterion for the prevention of statutory nuisances and for the control of contaminated land.

public liability policy

A policy of insurance covering a person or company in respect of claims made against them by third parties following injury, disease or damage to them or their property. Pollution claims under such policies can be problematic as public liability insurance generally only provides limited cover for the alleged polluter against sudden and unforeseen events causing a claim. Policies do not generally cover gradual pollution although cover may be available from specialist underwriters representative on the applicant's record of pollution control and total quality management. See also *insurance.*

pulverised fuel ash (PFA)

In the distillation of crude oil, pulverised fuel ash is the last component to be distilled at the highest temperature.

punitive damages

A pecuniary award made by a court inflicting or intending to inflict punishment on one person for the actionable wrong done by that person to another (more available in the US than in the UK). Such damages may be awarded with the intention of making an example of the defendant so that others are deterred from incurring similar avoidable liability. See also *Exxon Valdez.*

putrescible wastes

Wastes which consist mainly of animal or plant residues and which undergo degradation by bacteriological action, eg fruit and vegetable wastes, fats and fish residues.

pyrolysis

Destructive distillation through the reforming of material in an inert atmosphere to produce combustible gases, volatile fluids, tar and charcoal. Solid domestic refuse or municipal solid waste (MSW) refuse is placed in a retort and heated without additional air to temperatures between 500° C and 1000° C. The weight should be reduced by about 90% and the solid residue remaining is then disposed of at controlled landfill sites. Industrial waste can be reduced by up to 60% in volume.

quality assurance

All those planned and systematic actions necessary to provide adequate confidence that a product or service will satisfy given requirements for quality. See also *British Standard 5750.*

quality control

The operational techniques and activities that are used to fulfil requirements for quality. See also *British Standard 5750.*

quality management

That aspect of the overall management function that determines and implements the quality policy. See also *British Standard 5750.*

quality system

The organisational structure, responsibilities, procedures, processes and resources for the implementation of quality management. See also *environmental management system.*

Radioactive Incident Monitoring Network (RIMNET)

A system of national monitoring sites introduced following the Chernobyl incident to monitor radiation. Information from the network is provided to local authorities in relation to their emergency planning function.

radioactive material

Under the Radioactive Substances Act 1993, s.1, radioactive material includes substances which contain elements such as actinium, lead, polonium, proactinium, radium, thinium or uranium. Such elements may be present in the form of solids, liquids, gas or vapour.

Radioactive material also includes substances possessing radioactivity which is wholly or partly attributable to a process of nuclear fission or from a substance bombarded by neutrons or ionising radiations. This latter category does not, however, include substances possessing radioactivity as a result of processes occurring naturally, as a consequence of the disposal of radioactive waste, or as a result of contamination in the course of the application of a process to some other substance. It is an offence under s.6 of the Act for any unregistered person, with certain limited exceptions, to use radioactive material.

Radioactive Substances Act 1993 (RSA)

Came into force on August 27, 1993 and applies to the UK as a whole. The RSA is a consolidation of previous legislation, in particular the Radioactive Substances Act 1960, and prohibits the use of radioactive material and mobile radioactive apparatus together with the disposal and accumulation of radioactive waste without authorisation. The RSA is administered centrally by the Secretary of State for the Environment and by Inspectors appointed by him. The inspection of sites licensed under the Nuclear Installations Act 1965 is the responsibility of the Minister of Agriculture, Fisheries and Food. Separate arrangements apply to Northern Ireland. The RSA provides for the cancellation and variation of authorisations, enforcement and prosecution of offences.

radioactive waste

Waste *(qv)* which includes:

1 substances which if they were not waste would be radioactive material, and

2 a substance or article which has been contaminated in the course of production, keeping or use of radioactive material, or by contact with or proximity to radioactive waste (Radioactive Substances Act 1993, s.2).

It is an offence under s.13 to dispose of radioactive waste otherwise than in the terms of an authorisation. The disposal of radioactive wastes poses great technical difficulties for the nuclear industry, involving not only the safe deposit of waste but also the decommissioning and sealing of old power stations and managing and engineering safely the spent fuel. It looks likely that many of the nuclear power stations will not be dismantled until 100 years after they have stopped generating. Radioactive material produced today will still be dangerous in several million years, causing living cells to mutate or die.

The Government's current policies in relation to disposal of radioactive waste are spelt out in a 1994 Consultation Paper, *Review of Radioactive Waste Management Policy – Preliminary Conclusions*, which reviewed policy changes since the publication of the *National Strategy for Radioactive Waste Management* in 1984. The consultation paper highlights the following points:

1 the necessity for the reformulation of the present categorisation in terms of half life and activity, or other systems;

2 a regulatory framework should be established by the Government to protect the interest of the public, both now and in the future, for the minimisation of the creation, handling and treatment of nuclear waste, with no jeopardy to the safety of the workforce;

3 waste producers should develop their own waste management strategies, so that no waste management problems are created for which there are no identified solutions;

4 wastes are characterised in accordance with the physical and chemical properties to facilitate safe management and disposal;

5 progress of the disposal of waste accumulated at nuclear sites with appropriate timescale to be established;

6 management disposal costs for radioactive wastes should be formulated in accordance with the polluter pays principle;

7 high level wastes should be stored in a vitrified form for a minimum of 50 years to permit cooling.

See *House of Commons Environment Committee Report on Radioactive Waste* in 2 volumes, HMSO 1986: *Radioactive Waste* (HC 191) (Government reply published as Cmnd 9852).

radioactivity
The spontaneous emission of ionising particles and rays following the disintegration of certain atomic nuclei. Radioactivity can ooze from rock and soil, water and air but higher background levels are to be expected in areas containing deposits of uranium, thorium and radium.

radon
A naturally occurring radioactive gas produced from the radioactive decay of uranium, found in small quantities everywhere, but especially in areas of granite rock. Radon disperses quickly in the open, but can accumulate inside buildings. These accumulations can cause damage to the lungs and increase the risk of lung cancer. Recent estimates suggest that as many as one in 20 lung cancer cases in Britain might be caused by exposure to radon in the home with perhaps the greater incidence in areas such as Devon, Cornwall and Aberdeen. Further reference: House of Commons Environment Committee, *Indoor Pollution* (1992).

rainforest

A dense, luxuriant, closed mesomorphic community with global vegetation types containing many tree species associated with high rainfall and humidity and a relative absence of frosts. The distribution depends upon altitude, moisture availability, the nature of the soil, the aspect of the slope, the incidence of fire, damage by tropical cyclones and the impact of man. There are many types of rainforest although it is usual to consider three main divisions, the tropical, subtropical and temperate. See also *Rio Summit.*

Ramsar Convention

The Convention on Wetlands of International Importance Especially as Waterfowl Habitat, which was signed at Ramsar in Iran on February 2, 1971, and came into force on December 21, 1975. The United Kingdom became a party in 1976. The Convention does not permit regional economic co-ordination organisations to be signatories, so the EC is not a party to it. The Ramsar Convention seeks to promote the conservation of listed wetlands and their wise use, and requires States to designate alternative sites of the original habitat type should the development of any presently designated site become necessary in the urgent national interest. Sites are designated unilaterally by individual States. A site may be selected for the list of sites of international importance on the basis of its ecology, botany, zoology, limnology or hydrology, although the initial priority was for wetlands of international importance to wildfowl at any season. In the United Kingdom Ramsar List sites are protected by designation as sites of special scientific interest (SSSIs) *(qv)*. To date 47 sites have been identified in England, and are listed in Annex B to PPG9, *Nature Conservation* (1994). The Convention itself is reproduced as Annex E to PPG9, *Nature Conservation* (1994).

reclamation

The process of remanufacturing waste back into a new material, as opposed to recycling *(qv)* which is collecting, sorting and initial processing for further use. See also *Waste Framework Directive.*

recovery

The act or process of recovering or bringing back something to its normal state, eg the repairing of a car. This is distinguishable from recycling and reclamation. See also *Waste Framework Directive.*

recycling

Recycling includes processes aimed at making waste *(qv)* re-usable or reclaiming substances from it. Powers for recycling waste are given to waste disposal authorities (WDAs) *(qv)* and waste collection authorities (WCAs) *(qv)* by the Environmental Protection Act 1990, s.55 (for Scotland it is s.56). WDAs may make arrangements with waste disposal contractors *(qv)* for recycling waste, or for them to use waste to produce heat or electricity or both. Authorities may also buy waste for recycling, sell or otherwise dispose of waste or anything produced from waste. In Scotland these powers are limited to WDAs only. The Government's 1990 White Paper, *This Common Inheritance* (HMSO 1990) set a target of recycling 25% of household waste by the year 2000; current recycling rates are around 5% (though 26% in the case of glass).

recycling credits
Payments made for recycling and disposal of waste under the Environmental Protection Act 1990 (EPA 1990), s.52. Where a waste collection authority (WCA) retains waste for recycling, the waste disposal authority (WDA) is required to pay it a sum equal to the net saving to it of not having to collect it. Additionally, WCAs have a discretion to make payments to third parties engaged in recycling in their area on a similar basis. The Secretary of State for the Environment has power by regulations to make such payments compulsory in such circumstances as he may define. No such regulations have, however, been made to date. Net savings of expenditure are the amounts saved by the waste disposal or collection authority, as the case may be, through not having to dispose of or collect the waste in question. Any dispute over the calculation of these latter payments is to be resolved through arbitration. The Secretary of State has power to make regulations to assist in determining the net saving of expenditure and has made the Environmental Protection (Waste Recycling Payments) Regulations 1992 (SI 1992 No 462).

red (data) book
A catalogue published by the International Union for the Conservation of Nature, listing species which are rare or in danger of becoming extinct globally or nationally. Some catalogues are published by national authorities and species are included for which the national authority hosts a large part of the world's population and has an international responsibility for conservation.

red list
A list, resulting originally from the North Sea Conference of maritime nations in 1987, of 23 potentially dangerous substances that are toxic, persistent and liable to bio-accumulate arising from land-based sources. They appear in the Trade Effluent (Prescribed Processes and Substances) Regulations 1989, (SI 1989 No 1156), and are subject to

control in respect of discharge into sewers under integrated pollution control *(qv)* as special category effluent *(qv)* under the Water Industry Act 1991, s.138. They are the UK equivalent of black list substances prescribed under EC Directive 76/464 on pollution caused by certain dangerous substances discharged into the aquatic environment. They include mercury and its compounds, cadmium and its compounds, gamma-hexachlorocyclohexane, dichloro diphenyl trichloroethane (DDT), pentachlorophenol and its compounds, hexacholorobenzene, hexacholorobutadiene, aldrin, dieldrin, endrin, carbon tetrachloride, polychlorinated biphenyls, dichlorvos, 1, 2-dicholoroethane, tricholorobenzene, atrazine, simazine, tributyltin compounds, triphenyltin compounds, trifluralin, finitrothion, azinphos-methyl, malathion and endosulfan.

refuse derived fuel (RDF)

Fuel produced from domestic refuse, after glass and metals have been removed from it, by compressing it to form briquettes used to fuel boilers.

registered carrier

A person registered under the Controlled Waste (Registration of Carriers and Seizure of Vehicles) Regulations 1985. The Control of Pollution (Amendment) Act 1989, s.1 makes it an offence for any person who is not a registered carrier of controlled waste *(qv)* to transport controlled waste from one place to another in Great Britain, either in the course of business or otherwise for profit. A person guilty of an offence may on conviction by a Magistrates' Court be liable to pay a fine of up to £5,000. Local authorities, producers of controlled waste (other than building or demolition waste), British Rail, ferry operators, vessel operators licensed under the Food and Environmental Protection Act 1985, charities and voluntary organisations are exempt from registration. See also *duty of care as respects waste management and transportation.*

registers

Public registers of information are commonly used in the United Kingdom environmental legislation as a means of providing access to environmental information on matters such as applications for permits; decisions taken on applications; conditions imposed on the grant of a permit; service of enforcement and prohibition notices; revocation, discharge, variation or modification notices and other information.

Examples include those for: planning, enforcement and stop notices by local planning authorities *(qv)* under the Town and Country

Planning Act 1990; discharge consents *(qv)* by the National Rivers Authority (NRA) *(qv)* under the Water Resources Act 1991 (WRA 1991); trade effluent consents by sewerage undertakers *(qv)* under the Water Industry Act 1991; waste management licences by the WRA under the Environmental Protection Act 1990 (EPA 1990); integrated pollution consents by Her Majesty's Inspectorate of Pollution (HMIP) under the Part I EPA 1990; air pollution consents by the local authority under the EPA 1990. In some cases information may be withheld on grounds of commercial confidentiality, for example in connection with integrated pollution consents. Public access to the registers is as of right during normal office hours. See also *environmental information and observation network.*

regulation (EC)

An instrument of legislation of the European Communities (EC). By virtue of Article 189 of the Treaty of Rome, regulations have general application, and are binding in their entirety and directly applicable in all Member States. Thus no further legislative action is required by Member States to transpose the effect of regulations into domestic law. Generally use has been made of directives *(qv)* in environmental policy rather than regulations. Examples of EC regulations in the field of the environment include those on substances that deplete the ozone layer (91/594) on the eco-label award scheme (880/92), concerning export and import of certain dangerous chemicals (2455/92), and the supervision and control of shipments of waste within, into and out of the EC (259/93).

regulation (UK)

A form of subordinate legislation, made by statutory instrument and within powers conferred for the purpose by primary (ie statutory) legislation.

remedial action

The act of removing, counteracting, relieving, rectifying, or making good any damage which may have been caused, for example in the case of the environment by pollution.

remediation notice

A process proposed by the Environment Bill 1995 *(qv)* in relation to contaminated land. A remediation notice may require the carrying out of works to remedy the contamination or to assess the extent and nature of such contamination. Liability is imposed on the person who caused or permitted the contaminating substances to be there, but under certain circumstances passes to the owners and occupiers of the land for the time being. See also *contaminated land.*

renewable energy

The generic name to describe those energy flows which occur naturally and repeatedly in the environment, from the sun, the wind, the oceans, the fall of water, geothermal and biomass energy. These sources of energy have only recently assumed importance in Government environmental strategy, where their advantages are now seen not only in terms of their renewability, but also in reducing emissions of greenhouse gases *(qv)*. Some renewable sources produce no CO_2 at all. Wind supplied 1,000 megawatts of power in the EC in 1993, compared with just 37 megawatts in 1982. US wind farms are even more productive, supplying 1,700 megawatts. Solar power is expensive and is about 10 times the price of wind. It is expected to remain uneconomic for wide European use until 2010. In some cases, solar power is competitive for isolated sites far from electric grids, notably in the developing world. Until possibly fossil fuels are priced to include their full environmental costs, all forms of renewable energy will face difficulties in finding an adequate return on capital invested. Further reference: PPG22, *Renewable Energy* (1993). See also *wind energy*.

reprocessing

Can be a synonymous process with reclamation *(qv)* of substances for further use. More particularly, it is the word describing nuclear power fuel reprocessing, of which there are two facilities in the world, namely Thorp in Cumbria and Cogema in France, where spent fuel is separated into uranium, plutonium and waste product, behind thick concrete walls.

Twenty years ago, Thorp offered many advantages. The economic case assumed that 20,000 tonnes of spent fuel would be reprocessed in its first 10 years, which would ensure plant cost payback and profits for future investment. British Nuclear Fuels Ltd had little trouble in selling this capacity. Since then, BNFL has been seeking contracts for the second 10 years from 2004 to 2014. Until December 1984, 40% of this capacity, worth £9bn, has been sold. However, following the October 1994 German general election, there have been two cancellations totalling 450bn tonnes, cutting back capacity by 23%, with a £360bn loss of business, less £100m in cancellation penalties. These cancellations show that the tide can be turning against reprocessing. Since Thorp was conceived, uranium supplies have multiplied and plutonium has lost its military appeal.

Resource Conservation and Recovery Act 1976 (RCRA)

US Federal legislation which provides a cradle to grave waste management programme as a response to problems associated with hazardous waste disposal in landfills. RCRA provides for:

1 the identification and listing of hazardous wastes,

2 standards for treatment, storage and disposal facilities,

3 enforcement, and

4 insurance. RCRA applies to hazardous substances listed in regulations made under RCRA and to other solid wastes which possess dangerous characteristics. RCRA imposes a duty on waste generators to test unlisted waste for the presence of dangerous characteristics by reference to factors such as whether the waste contains listed substances, is acutely toxic, or likely to be ignitable, corrosive, reactive or toxic.

RCRA also requires that waste be documented during transport, that responsible carriers be employed, and that the final destination of the waste is a licensed landfill site. Standards for treatment, storage and disposal facilities are similarly stringent. Facility and site licences are required. Standards apply to discarding chemical products, use of underground storage tanks, and burning of fuel containing hazardous wastes. They apply to operations other than landfills alone, and include activities such as dry-cleaners and motor vehicle repair businesses. The objectives of the standards are:

1 to impose design and operation requirements to prevent releases of waste whilst the facility is in operation and for 30 years thereafter, and

2 performance criteria for groundwater protection.

The Environmental Protection Agency (EPA) is responsible for enforcing RCRA (unless states have introduced their own legislation, which most have). State legislation must not be less stringent than RCRA. The EPA has reserve enforcement powers if states do not act within 30 days. Both civil and criminal penalties are imposed. Temporary or permanent injunctions may be granted, or civil penalties of $25,000 per day may also be imposed for violation of administrative orders. Contravention of RCRA which causes imminent danger of death or serious bodily harm may result, on criminal conviction, in fines of $250,000 and up to 15 years imprisonment. Companies may be fined up to $1,000,000. Other contraventions may, following conviction, result in fines of $50,000 per day and imprisonment of between two and five years. RCRA also provides a parallel insurance safeguard to pay for clean-up and compensate for injuries as a back-up for cases where enforcement under the Act does not provide an effective remedy.

Resource Management Act 1991 (RMA) New Zealand

An innovative legislative measure which represents a new approach to environmental regulation. Its purpose is to combine the sustainable management of natural and physical resources and it embraces land-use planning, coastal policy, water resources, discharges into the environment and noise. It is administered by district and regional councils whilst the Minister for the Environment and Conservation occupies a supervisory role. Appeals and enforcement are determined by a specialist environmental court historically termed the Planning Tribunal. Other NZ environmental enactments such as the Historic Places Act 1993 make use of the provisions of the RMA in relation to consents, appeals and enforcement.

responsible care programme

A programme devised by the Chemical Industries Association *(qv)*, and composing guidelines for certification to ISO 9001, health and safety and environmental management systems and British Standard 7750 *(qv)* in the chemical industry. It is an umbrella programme, attaining worldwide acceptance, which is designed to improve the performance of the chemical industry in the fields of health, safety, environment, product safety, distribution, emergency response and relations with the public, and to enable companies to demonstrate that these improvements are, in fact, taking place.

Richter Scale

A measure of seismic activity that was devised by Charles Richter at the California Institute of Technology at Pasadena in 1935 to provide himself and his colleagues with a standard measure of earthquake strength. Seismology's advance, however, has refined, mutated and adapted Richter's idea almost beyond recognition. The scale in common use today is the seismic movement magnitude which measures the strength of the long period waves that accompany the most frantic shaking. They are reckoned by Richter's successors to be the best possible index of the energy released.

Ringelmann Charts

A series of six charts, numbered 0–5 and shaded from white to black, showing shades against which the density of smoke may be measured. The charts are used for the control of dark smoke under the Clean Air Act 1993 Part I. See also *smoke*.

Rio Summit (UNCED)

The UN Conference on Environment and Development which was held in Rio de Janeiro in 1992. The Rio Summit resulted in the conclusion of conventions on Climate Change and Biological Diversity, a statement of principles for the sustained management of forests, Agenda 21 (an environmental action plan for the next century) and the Rio Declaration of 27 principles on the environment and development. It also resulted in the establishment of the Sustainable Development Commission to monitor implementation of the conventions at national level.

The UN Framework Convention on Climate Change commits states to adopt programmes to contain greenhouse gas emissions, and in developed countries to reduce emissions (particularly of CO_2) to 1990 levels by the year 2000. The convention on Biological Diversity obliges states to identify and monitor their biological resources, and requires them to produce conservation plans and designate protected areas. Provision is also made for aid in the case of developing countries. Agenda 21 aims to ensure that environmental concerns are taken into account when decisions are taken on matters such as energy and agriculture. Provision is also made for aid and technology transfer to assist developing countries implement their action plans. The Rio Declaration puts forward principles designed to enhance citizen action on the environment, a market-based approach to environmental costs, the effectiveness of legislation and dispute resolution.

The UK Government published four White Papers on January 25, 1994 to show how they proposed to fulfil the obligations entered into at the Rio Summit. They set out Government policy on climate change, biological diversity, forests and sustainable development. The four White Papers are *Sustainable Development: The UK Strategy,* Cm 2426 (1994); *Climate Change: The UK Programme,* Cm 2427 (1994); *Biodiversity: The UK Action Plan,* Cm 2428 (1994) and *Sustainable Forestry: The UK Programme,* Cm 2429 (1994). See also *biodiversity; forestry.*

riparian rights

The rights enjoyed by the owners of land adjacent to a river, which includes rights of abstraction *(qv)* of water, fishing, and extraction of sand and gravel from the river bed and the right not to have water diminished in flow or quality.

risk

The likelihood of a potential hazard being realised. The extent of the risk is dependent both on the likelihood of something unsafe happening and a hazardous event occurring and the extent of damage it would cause if it did. See *Living With Risk, The British Medical Association Guide* (Wiley,1988); *Major Industrial Hazards: Their Appraisal & Control* (John Withers, Gower Press, 1988).

risk assessment

A technique used in assessing the relative costs and benefits of a particular situation or development. The assessment will compare the recorded incidence of damage or death or injury to the number of times that the subject of the assessment has been used over a reasoned length of time. The technique of risk assessment is usually reliable but takes no account of perceived risk *(qv)*.

River Invertebrate Prediction and Classification System (RIVPACS)

A biological index, developed by the Institute of Freshwater Ecology, to measure the degree of pollution of a waterbody by assigning scores to the presence of macro-invertebrate communities found at a given site with the community scores that would be predicted to occur under optimum conditions. Further reference: Royal Commission on Environmental Pollution *16th Report, Freshwater Quality* (Cm 1966; 1992).

River Purification Board (Scotland)

Established under the Local Government (Scotland) Act 1973 to promote the cleanliness of rivers, inland and tidal waters within their areas, to conserve water resources, and to exercise powers under the Rivers (Prevention of Pollution) (Scotland) Act 1951 and the Control of Pollution Act 1974, eg in respect of discharge consents, remedying pollution, dealing with waste from vessels, and prohibiting agricultural practices. There are seven River Purification Boards covering the areas of the Clyde, Firth, Highland, North East, Solway, Tay and Tweed. Beyond the mainland of Scotland the powers are exercised by the islands councils (together they are known as River Purification Authorities). Members of River Purification Boards comprise representatives of regional and district councils and persons appointed by the Secretary of State interested in matters such as agriculture, fisheries and industry. The Environment Bill 1995 proposes the transfer of the function of the River Purification Authorities to the Scottish Environment Protection Agency.

road

As defined by the Road Traffic Regulation Act 1984 (RTRA), s.142(1) includes 'any length of highway or any other road to which the public has access, and includes bridges over which a road passes'. The public must therefore either have access as of right, or access must be tolerated. Thus cases where a physical obstruction has to be overcome, or a prohibition defied in order to gain access, would not come within the definition in the RTRA 1984. Roads themselves comprise both the carriageway (normally used by vehicles) and the footway (used exclusively by pedestrians) (see *Bryant v Marx* [1932] All ER Rep 518).

Road Traffic (Carriage of Dangerous Substances in Road Tanker & Tank Containers) Regulations 1992 SI No 743

Regulations establishing the legal requirements applicable to the transport of hazardous substances in tankers and tank containers. The classes used are based on those defined in the United Nations Committee of Experts' publication *Recommendations on the Transport of Dangerous Goods* (the UN orange book).

Road Traffic Regulation Act 1984 (RTRA)

The RTRA provides for, *inter alia*, the making of Traffic Regulation Orders (TRO) by county councils in England and elsewhere by metropolitan districts. Separate arrangements apply in London. TROs may be used to regulate the use of roads by specified types of traffic, including through traffic, and also for specifying through routes for heavy commercial vehicles and prohibiting or restricting their use in specified roads or zones, so as to preserve or improve the amenities of the area.

rotary kiln

A kiln used principally for heating loose bulk materials such as cement and lime, but also used for other purposes such as the incineration of combustible waste. It is a long cylinder supported on steel hoops and riding on rollers. This is inclined at a slight slope and rotated at a slow speed. A burner is placed at the lower end of the kiln, while the feed enters at the upper end and falls through the kiln, being heated by the upward flow of hot combustion gases.

Round Table on Sustainable Development

A committee proposed by the British Prime Minister in January 1994 as one of three mechanisms for carrying forward the country's programme for sustainable development *(qv)* (see also *Government Panel on Sustainable Development* and 'Going for Green'). The Round Table has a membership of 30, drawn from business, non-governmental organisations (NGOs) *(qv)*, local government, consumers, trades unions, medicine, agriculture and other groups. Its objectives are to develop consensus on approaches to sustainable development, to help identify the agenda and priorities for sustainable development, to provide recommendations and advice on action to achieve and to help evaluate progress.

Royal Commission on Environmental Pollution (RCEP)

A standing commission appointed by the Government by Royal Warrant in 1970, 'to advise on matters, both national and international, concerning the pollution of the environment; on the adequacy of research in this field; and the future possibilities of danger to the environment'. The Royal Commission sets its own agenda, and prepares major reports based upon research and upon evidence supported to it. Its reports are presented to Parliament as command papers. Those published to date are:

1st Report Cmnd 4585 (1971)

2nd Report:
Three Issues in Environmental Pollution, Cmnd 4894 (1972)

3rd Report:
Pollution in some British Estuaries and Coastal Waters, Cmnd 5054 (1973)

4th Report:
Pollution Control: Progress and Problems, Cmnd 5780 (1974)

5th Report:
Air Pollution – an Integrated Approach, Cmnd 6371 (1976)

6th Report:
Nuclear Power and the Environment, Cmnd 6618 (1976)

7th Report:
Agriculture and Pollution, Cmnd 7644 (1979)

8th Report:
Oil Pollution and the Sea, Cmnd 8358 (1981)

9th Report:
Lead in the Environment, Cmnd 8852 (1983)

10th Report:
Tackling Pollution – Experience and Prospects, Cmnd 9149 (1984)

11th Report:
Managing Waste: the Duty of Care, Cmnd 9675 (1985)

12th Report:
Best Practicable Environmental Option, Cm 310 (1988)

13th Report:
The Release of Genetically Engineered Organisms to the Environment, Cm 720 (1989)

14th Report:
GENHAZ: A System for the Critical Appraisal of Proposals to Release Genetically Modified Organisms into the Environment, Cm 1557 (1991)

15th Report:
Emissions from Heavy Duty Diesel Vehicles, Cm 1631 (1991)

16th Report:
Freshwater Quality, Cm 1966 (1992)

17th Report:
Incineration of Waste, Cm 2181 (1993)

18th Report:
Transport and the Environment, Cm (1994)

Royal Society for Nature Conservation
A UK non-governmental organisation for nature conservation which has been involved in the Government's biodiversity programme and has also organised a wild life trusts partnership. [*Address: The Green, Witham Park, Waterside, Lincs, LN5 7JR; T: 01522 544 400.*]

Royal Society for the Protection of Birds (RSPB)
A non-governmental organisation dedicated to the conservation of birds. It has over 860,000 members (which makes it the largest wildlife conservation charity in Europe), and manages over 130 nature reserves in the United Kingdom (over 83,000 hectares). The RSPB manages about 5% of Britain's lowland grassland and intertidal areas, which provide refuges for a quarter of the United Kingdom's wildfowl and wading bird populations in mid-winter. [*Address: The Lodge, Sandy, Beds, SG19 2DL: T:01767 680551; F: 01767 692365.*]

run-off
The discharge of water from land through surface flows into larger watercourses.

Rylands v Fletcher
A 19th-century law case with important implications for contemporary environmental liability *(qv)* in which it was established that strict liability *(qv)* applied to a person who brings onto his land something which is likely to cause damage if it escapes, for any damage caused irrespective of fault: 'We think that the true rule of law is, that the person who for his own purposes brings on his lands and collects and keeps there anything likely to do mischief if it escapes, must keep it in at his peril, and, if he does not do so, is

prima facie answerable for all the damage which is the natural consequences of its escape' per Blackburn, J. (1866) LR 1 EX 265 at 279–280. An attempt to extend this doctrine to environmental damage from the escape of hazardous substances was partially successful in the House of Lords in the *Cambridge Water Company* case *(qv)*, where it was held that the storage of substantial quantities of chemicals on industrial premises should be regarded as an almost classic case of non-natural use, but that foreseeability of damage of the relevant type was a prerequisite of liability in damages not only in nuisance, and negligence, but also under the rule in *Rylands v Fletcher*.

salami effect

The process of slicing a major project into small parts and seeking regulatory approval for each part, so as to remain under the threshold for special regulatory requirements, such as environmental assessment. The High Court has held that it is still incumbent on the decision-maker to have regard to the overall project, and not just the part, when deciding whether environmental assessment is necessary: *R v Swale Borough Council, ex parte RSPB* (1990) *Journal of Environmental Law* Vol 3; 135.

Sandoz

The major industrial pollution incident that occurred at the Sandoz Chemical plant in Basle, Switzerland, in 1986, when run-off from the water used in fire-fighting carried between 13 and 30 tonnes of chemicals into the River Rhine, causing serious damage to flora and fauna, and polluting the underlying aquifer *(qv)*.

saturated zone

The zone of porous and/or fractured geological strata in which all voids are filled with water. The upper boundary is the water table *(qv)* above which pores and/or fissures are partly filled with air, and which is referred to as the unsaturated zone.

Schedule 9 obligation

The environmental obligation imposed on electricity suppliers and distributors (such as the National Grid and Regional Electricity Companies) by Schedule 9 of the Electricity Act 1989, when formulating proposals for large generating stations, for installing electric lines and for other works relating to the transmission or generation of electricity. Schedule 9 obliges them to have regard to the desirability of preserving natural beauty, of conserving flora, fauna and geological and physiographical features of special interest, and of protecting sites, buildings and objects of architectural, historic or archaeological interest, and to do what they reasonably can to mitigate any effect which their proposals might have on the natural beauty of the countryside or on such flora, fauna, features, sites, buidings or objects. Moreover, the Schedule requires them to publish a statement of how they propose to meet this responsibility.

Scottish Environment Protection Agency (SEPA)

A new agency proposed for Scotland under the Environment Bill 1995, to which will be assigned the current functions of the river purification authorities, waste regulation authorities, waste disposal authorities and miscellaneous environmental protection functions. The Secretary of State for Scotland is required to give guidance to SEPA, including guidance as to the contribution he considers it appropriate for SEPA to make towards attaining the objective of sustainable development *(qv)*.

Scottish Natural Heritage (SNH)

A statutory body, formed in 1992, which has the responsibility for nature conservation matters in Scotland. Its functions were previously undertaken by the Nature Conservancy Council, its short-lived successor the Nature Conservancy Council for Scotland and the Countryside Commission for Scotland. SNH is responsible for the designation and management of nature reserves, for the designation of sites of special scientific interest, for some other forms of habitat protection, and for licensing activities otherwise unlawful under the laws protecting individual species.

Scottish Office Agriculture and Fisheries Department (SOAFD)

The Department responsible for the implementation of the Common Agricultural Policy and Fisheries Policy in Scotland as well as a wide range of environmental functions, including Environmentally Sensitive Area Schemes (ESAS). See also *Ministry of Agriculture, Fisheries and Food.*

Scottish Office Environment Department (SOEnD)

The Department responsible for implementing for Scotland Government policy on environmental protection, town and country planning, and countryside conservation. It currently has two divisions dealing with pollution control, namely Her Majesty's Industrial Pollution Inspectorate *(qv)* and the Hazardous Waste Inspectorate. It also has an appellate function in respect of disputed matters referred from other regulatory bodies.

Scottish Renewables Order (SRO)

The equivalent in Scotland (operated by the Secretary of State for Scotland) to the non-fossil fuel obligation *(qv)*.

scrap metal

Any old metal, and any broken, worn out, defaced or partly manufactured articles made wholly or partly of metal and any metallic wastes; also includes old, broken, worn out or defaced tooltips or dies made of any of the materials commonly known as hard metals or of cemented or sintered metallic carbides (Scrap Metal Dealers Act 1964, s.9(2)). Metal recycling sites must be registered and be subject to inspection by the Waste Regulation Authority *(qv)* but they are not subject to the requirement of being managed by a fit and proper person *(qv)* to hold a waste management licence. This regulatory regime may not reflect the real risks and may be reviewed. See also *controlled waste; recovery; recycling.*

scrubber

Gas absorption apparatus used for removing impurities from gas or vapour. Often used in connection with flue gas desulphurisation *(qv)* in coal-fired electricity generating stations, but also relevant to air pollution control in other industrial activities emitting gases to the atmosphere.

sensitive waters

A designation by the Secretary of State under EC Directive 91/271 on urban waste water treatment for the protection principally of natural freshwater lakes and other waters which are or may become liable to eutrophication *(qv)*, and for surface fresh water used for potable water supply and which could contain higher than permitted concentrations of nitrate. Under the Directive, discharges from sewage treatment works will be permitted to such waters only where they have received tertiary treatment, and meet concentration limits for phosphate and/or nitrate. See also *less sensitive waters; normal waters; Urban Waste Water Treatment Directive.*

several liability

See *joint and several liability.*

Seveso

A town near Milan which gave its name to an important EC Directive after the escape, in 1976, of toxic dust containing dioxins *(qv)*, from a factory located there which spread to the surrounding countryside causing widespread injury and damage but most fortunately no fatalities. The disaster prompted EC Directive 82/501 on the major accident hazards of certain industrial activities, which is concerned with exceptional risks posed by industrial activity such as fires, explosions and significant emissions of dangerous substances when an industrial activity gets out of control. The Directive places a general duty on manufacturers using a wide range of processes to prevent major accidents and to limit their consequences for man and the environment (Articles 3 and 4); there is also a general duty to report major accidents. The Seveso Directive was transposed in the UK by the Control of Major Accident Hazards Regulations 1984 *(qv)*. The incident at Seveso provides one of the few sources of data on human exposure to high levels of dioxins *(qv)*.

sewage

Waste, especially excremental matter, conveyed in sewers. Sewage effluent includes any effluent from the sewage disposal or sewerage works of a sewerage undertaker *(qv)*. See also *trade effluent; Urban Waste Water Treatment Directive.*

sewage sludge

The accumulated solids produced during the treatment of sewage. Sewage sludge contains heavy metals, and can also release pathogens and bacteria into the air which could cause harm to human health. The principal methods of disposal have been:

1 tipping at sea, which the United Kingdom Government is now pledged to phase out by the end of 1998 under the North Sea Conference and the EC urban waste water Directive;

2 incineration;

3 landfill *(qv)*; and

4 spreading on agricultural land. Its use on agricultural land must be regulated to ensure that heavy metal accumulation in the soil does not exceed certain limits, and is governed by EC Directive 86/278 in an attempt to ensure that human beings, animals, plants and the environment are fully safeguarded against the possibility of harmful effects from the uncontrolled spreading of sewage sludge on agricultural land; and to promote the correct use of sewage sludge on such land. The Directive was implemented in the UK by the Sludge (Use in Agriculture) Regulations 1989 (SI 1989 No 1263).

Further reference: House of Lords Select Committee on the European Communities, *Sewage Sludge in Agriculture* (1983) HMSO.

sewer

A channel or conduit, now usually covered and underground, built to carry off and discharge drainage water and sewage. The Water Industry Act 1991, s.219(1) defines a sewer as including all sewers and drains used for the drainage of buildings and yards appurtenant to buildings but does not include drains which are used for the drainage of one building or of any buildings or yards appurtenant to buildings within the same curtilage.

sewerage

Drainage by sewers: a system of draining by sewers. Sewerage services are defined in s.219(1) of the Water Industry Act 1991 as including the disposal of sewage and any other services which are required to be provided by a sewerage undertaker for the purposes of carrying out its function. The Public Health Act 1875 made it the duty of local authorities to deal effectively with the drainage of their districts and gave a right to owners and occupiers of premises within the district of a local authority to drain their premises into the public sewer, thereby laying the foundations of a drainage and sewerage system which lasted until the Water Act 1989, which privatised the water industry. The responsibility today rests with the sewerage undertakers *(qv)*.

sewerage undertaker

A company holding appointment under the Water Industry Act 1991 to provide public sewerage services in a specified area. Ten companies were appointed in 1989, as successors to the former regional water authorities, to act as both sewerage and water undertakers *(qv)* (sometimes known as water and sewerage plcs), although in some parts of their areas public water supply responsibilities may be the function of statutory water companies. Further reference: Department of the Environment (DOE) and Welsh Office (WO), *Instrument of Appointment of the Water and Sewerage Undertakers* (HMSO, 1989).

shadow director

A person in accordance with whose directions or instructions the directors of a company are accustomed to act (Companies Act 1985 (s.741(2)). However a person is not deemed a shadow director by reason only that the directors act on advice given by him in a professional capacity. Many of the statutory rules which are applicable to directors also apply to shadow directors.

shellfish waters

Waters in which shellfish grow, and which are therefore entitled to protection under EC Directive 79/923, which required Member States to designate coastal and brackish waters which needed protection or improvement to support shellfish, and to establish pollution reduction programmes to ensure that all designated waters complied within six years with the specified quality parameters. Further reference: National Rivers Authority, *Implementation of the EC Shellfish Waters Directive* (Water Quality Series No 16; 1994).

short-term exposure limit

A limit prescribed by the Health and Safety Regulations, such as the Control of Substances Hazardous to Health Regulations 1988 (COSHH) *(qv)* and their Approved Codes of Practice, defining the maximum time for which an employee or a third party can be exposed to a particular substance. An employer must not carry on any work which is liable to expose any employee to any substance hazardous to health unless he has made a suitable and sufficient assessment of the risks created by that work to the health of that employee and of the steps that need to be taken.

sick building syndrome

The name given to the condition in which the occupants of buildings experience symptoms, such as headaches, dry skin, dry throat, stuffy nose, lethargy, watering eyes and lack of concentration, which disappear soon after they leave the building. In some cases there are clear causes, such as poor ventilation, heating or lighting systems. In others, there may be no single cause. The significance of the syndrome is a matter of some dispute. Research into it has been undertaken by the World Health Organisation and by the UK Building Research Establishment but, because it is not life-threatening, it has not been seen by the Health and Safety Executive as being of high significance. Further reference: House of Commons Environment Committee, *6th Report (Session 1990–91) Indoor Pollution* (HC61; 1991).

silage

Green fodder preserved by pressure in a silo or a stack.

silviculture

The cultivation of woods or forests; the growing and tending of trees as a branch of forestry *(qv)*. See also *afforestation.*

site of community importance

A site identified by the Secretary of State under the Conservation (Natural Habitats, &c.) Regulations 1994 (SI 1994 No 2716), reg. 7 in accordance with EC Directive 92/42 (the Habitats Directive *(qv)*),

Article 4. Once identified, such a site must be designated as a special area of conservation (SAC) *(qv)*. See also *European Site.*

site of special scientific interest (SSSI)

An area of land notified under the Wildlife and Countryside Act 1981 by the appropriate nature conservation body (in England, the Nature Conservancy Council; in Wales, the Countryside Council for Wales; in Scotland, Scottish National Heritage) as being of special interest by virtue of its flora and fauna, geological or physiographical (landform) features.

Owners and occupiers of SSSIs must give the appropriate body four months' written notice of intention to carry out any operation listed in the notification as likely to damage the special interest of the site. This allows time for the appropriate body to consider the implications of the proposal and to discuss any modifications which would avoid damage to the wildlife or geological interest of the site, and they may invite the owner or occupier to enter into a management agreement for the site in return for payment. If no agreement is concluded, there is no impediment to the works being carried out unless the Secretary of State proceeds to make a nature conservation order *(qv)* (under s.29 of the Wildlife and Countryside Act 1981) which extends the statutory negotiation period. Compulsory purchase powers are also available. There is a duty on local planning authorities to consult with the appropriate body on any planning applications for development in, or likely to affect, an SSSI under the Town and Country Planning, General Development (Procedure) Order 1995. It is intended that all national nature reserves *(qv)*, terrestrial Ramsar *(qv)* sites, special protection areas *(qv)* and special areas of conservation *(qv)* should first be notified as SSSIs.

Protection given by legislation is generally regarded as weak and has been described judicially by the House of Lords as 'toothless' (see *Southern Water Authority v Nature Conservancy Council* [1992] 3 All ER 48. Further reference: PPG9, *Nature Conservation* (HMSO, 1994); National Audit Office, *Protecting and Managing Sites of Special Scientific Interest in England* (HMSO, 1994).

Sizewell

A site on the Suffolk coast where the Sizewell B pressurised water reactor (PWR) was constructed between 1980 and 1994. The process of authorisation included a public local inquiry which lasted a record 340 days. The major area of debate was the safety of the technology, which had been employed in the US but not previously in the UK. The reactor was commissioned in 1995, and application has since been lodged for approval of a third nuclear reactor (Sizewell C) at the same site.

slurry

Thin semi-liquid mud or cement. Also any semi-liquid mixture of pulverised solid or fine particles with a liquid; farmyard manure in fluid form. The escape of farmyard slurry is a common cause of watercourse pollution, and minimum standards for the slurry storage systems are prescribed by the Control of Pollution (Silage, Slurry and Agricultural Fuel Oil) Regulations 1991 (SI 1991 No 324), Sched 2. Compliance with the requirements is not a defence to prosecution for water pollution but may be a mitigating factor. Further reference: National Rivers Authority, *Water Pollution Incidents in England and Wales – 1993* (NRA Water Quality Series No 21: September, 1994).

smells

Odours, produced by small concentrations of organic vapours which can produce unpleasant reactions in humans and animals and perhaps plants.

smoke

The visible suspension of carbonaceous particulates greater than one micron in diameter, and other particles in air, given off by a burning or smouldering substance. The Clean Air Act 1993 restricts the emission of dark smoke from any chimney, and the emission of dark smoke from any industrial or trade premises whether or not from a chimney. The 1993 Act defines dark smoke to be smoke which is as dark as or darker than shade 2 on the Ringelmann Chart. Ringelmann Charts are pieces of card which when viewed at an appropriate distance appear to show a variety of shades from black to white. Smoke is also defined in s.79(7) of the Environmental Protection Act 1990 to include 'soot, ash, grit and gritty particles emitted in smoke'.

solar energy

Energy obtained from the sun's rays. Detailed advice on the exploitation of solar energy and its land-use implications is contained in a special Annex issued in 1994 to PPG22, *Renewable Energy* (1993).

solvent

A liquid which has the capability to dissolve other substances.

solvent extraction

The partial removal of a substance from a solution or a mixture of liquids by utilising its greater solubility in another liquid.

sound wave

Pressure variations in the air caused by a series of compressions and rarefactions moving outwards from some vibrating object. The frequency of the sound wave is the rate at which successive pressure variations reach the eardrum. Each wave, first a compression then a rarefaction, on encountering an object exerts a force. This is why sound can break a glass or vibrate a window. Frequency is expressed as the number of vibrations per second or cycles per second. One cycle per second equals one hertz (Hz). The frequency of a wave is also the number of complete waves that pass a given point in one second.

source protection zone

A zone designated by the National Rivers Authority *(qv)* with the aim of protecting aquifers *(qv)* from over-abstraction and pollution. There are three zones:

Zone I (inner source protection) is located immediately alongside the groundwater source (ie a spring, well or other abstraction point) and is defined by a 50-day travel-point from any point below the water table to the source and as a minimum of 50 metres radius from the source);

Zone II (outer source protection, defined by a 400-day travel time);

Zone III (source catchment, which covers the complete catchment area of a groundwater source).

The boundaries of the zones are defined in light of best information and technical advice available at the time, but are subject to regular reappraisal in light of new knowledge or changed circumstances. The zones provide a basis for decisions by the Authority on applications for abstraction licences and discharge consents, and also for consultation with local planning authorities in relation to the location of new development and waste regulation authorities in relation to licensing of landfills. For example, the Authority will normally object to all activities requiring a waste management licence within an area designated as Zone I. Further reference: National Rivers Authority, *Policy and Practice for the Protection of Groundwater* (1992).

special area of conservation (SAC)

An area designated under EC Directive 92/43 on the conservation of natural habitats and of wild fauna and flora (the Habitats Directive *(qv)*). The Directive requires Member States to take measures to maintain or restore natural habitats and wild species at a favourable conservation status in the Communities, giving effect to both site and species protection objectives. Following a period of consultation, sites to be designated as special areas of conservation must be agreed with the EC Commission by June 1998.

special category effluent

Trade effluent *(qv)* which contains substances prescribed by the Secretary of State, or which derives from some prescribed process (Water Industry Act 1991, s.138), and which is therefore subject to a special set of controls in relation to its discharge into public sewers. See also *red list.*

special industrial uses

Uses of land that were within any one of four use classes (Classes B3 to B7) within the Town and Country Planning (Use Classes) Order 1987 *(qv)*, which were industrial uses which may cause nuisance or adversely affect the environment. The Order allowed a change to occur between different uses within the same class without planning control. The Government consulted on proposals to abolish them and to substitute two classes based upon the Environmental Protection (Prescribed Processes and Substances) Regulations 1991. The special classes were then repealed from March 9, 1995 by SI 1995 No 297.

special protection area (SPA)

Area so designated under EC Directive 79/409 on the conservation of wild birds (see *Birds Directive*), as amended by the Habitats Directive 92/43 *(qv)*. The Directive requires Member States to take measures for the conservation of certain wild birds, by protecting their habitat by classifying areas as special protection areas. Implementation of the Directive was originally primarily through the Wildlife and Countryside Act 1981, including the prohibitions on killing, taking or injuring birds, or destroying nests and eggs, in Part I; and the protection of special areas by designating SSSIs under Part II. Those powers are now supplemented by the Conservation (Natural Habitats, &c.) Regulations 1994 (SI 1994 No 2716). The Directive is currently being reviewed alongside proposals for the implementation of the Habitats Directive for which the protective criteria are now identical. See also *site of special scientific interest (SSSI).*

special waste

A term sometimes broadly used to cover toxic waste, hazardous waste and dangerous waste, but which has a specific definition in regulations made under the Environmental Protection Act 1990, s.75(9), and is subject to a special regime of controls. The Act requires the Secretary of State to make regulations governing the treatment of 'controlled waste [*qv*] of any kind [which] is or may be so dangerous or difficult to treat, keep or dispose of that special provision is required for dealing with it'. The current regulations dealing with special waste are the Control of Pollution (Special Waste) Regulations 1980 (SI 1980 No 1709) as amended by SI 1988 No 1790. These Regulations, together with others, transpose EC Directive 78/319 on toxic and dangerous waste. The Government is currently reviewing the Special Waste Regulations in the light of two important developments:

1 EC Directive 91/689 on hazardous waste, which includes a new definition superseding that of Directive 78/319; and

2 the Basle Convention *(qv)* on the Control of Transboundary Movements of Hazardous Wastes and their Disposal, which has now entered into force. A Consultation Paper, *Special Waste and the Control of its Disposal*, was issued by the Department of the Environment and Welsh Office in 1990 but will require updating in light of the EC Directive.

See also *hazardous (toxic) waste.*

springs

Sources of surface water where the groundwater or the water table flows under natural pressure to the surface.

statutory instrument (SI)

The form in which most subordinate legislation is now made, and which is governed by the Statutory Instruments Act 1946. Statutory instruments must be laid before Parliament and, according to the statute under which they are made, may require a positive resolution of both Houses in order to take effect (affirmative resolution procedure) or will take effect within 21 days of being laid unless either House resolves otherwise (negative resolution procedure). Extensive powers to regulate by statutory instrument are conferred by environmental legislation, and also by the European Communities Act 1972, s.2, which allows the Government to transpose European legislation by statutory instrument rather than having to secure enactment of primary legislation.

statutory nuisance

A statutory nuisance is an activity which, whether or not it constitutes a nuisance at common law, is made a nuisance by statute (either in express terms or by implication). The purpose is to establish a summary procedure to secure their abatement *(qv)*. Under the Environmental Protection Act 1990 s.79(1), the following matters constitute statutory nuisances:

1 any premises in such a state as to be prejudicial to health or a nuisance;

2 smoke emitted from premises so as to be prejudicial to health or a nuisance;

3 fumes or gases emitted from premises so as to be prejudicial to health or a nuisance;

4 any dust, steam, smell or other effluvia arising on industrial, trade or business premises and being prejudicial to health or a nuisance;

5 any accumulation or deposit which is prejudicial to health or a nuisance;

6 any animal kept in such place or manner as to be prejudicial to health or a nuisance;

7 noise emitted from premises so as to be prejudicial to health or a nuisance;

7a noise that is prejudicial to health or a nuisance and is omitted from or caused by a vehicle, machinery or equipment in the street;

8 any other matter declared by any enactment to be a statutory nuisance.

The Act imposes a duty on every local authority to cause its area to be inspected from time to time to detect any statutory nuisances and, where complaint of a statutory nuisance is made to it by a person living within its area, to take such steps as are reasonably practicable to investigate the complaint. Where the local authority is satisfied that a statutory nuisance exists, or is likely to occur or recur, they must serve an abatement notice *(qv)* on the person responsible for the nuisance. The notice must identify clearly and precisely the nuisance complained of and tell the recipient what is required of him, that is, what works or steps are necessary to abate or prevent the recurrence of the nuisance. The notice must include a statement informing the recipient of his right to appeal under s.80(3) and giving the time limit for such appeal.

statutory undertaker

A body carrying out a quasi-public function for which it is specifically authorised by statute, and for which it enjoys certain privileges, such as rights of entry to privately owned land, special permitted development rights or powers of compulsory purchase of land. A typical definition is that provided by the Town and Country Planning Act 1990, s.262, as meaning any person authorised by any enactment to carry on any railway, light railway, tramway, road transport, water transport, canal, inland navigation, dock, harbour, pier or lighthouse undertaking. It also includes, for some purposes, any water or sewerage undertaker *(qv)*, and the National Rivers Authority *(qv)*.

statutory water quality objective (SWQO)

See *water quality objective*.

stochastic

Randomly determined; that which follows some random probability distribution or pattern, so that its behaviour may be analysed statistically but not predicted precisely. Having a probability distribution, usually with finite variance.

storm overflow

A device on a combined or partially separate sewerage system introduced for the purpose of relieving the system of excessive flows, with the excess flow being discharged to a convenient watercourse.

stratification

The action of depositing something in layers. The formation, by natural process, of strata or layers one above the other. The fact or state of existing in the form of strata; the manner in which something is stratified. Types include:

1 geological – the formation of strata in portions of the crust of the Earth by successive deposits of sedimentary matter; the manner in which a portion of the crust of the Earth is stratified. (Each layer represents the sediment deposited over a specific period.)

2 biological and pathological – the thickening of a tissue by the deposition or growth of successive layers.

3 electrical – the striated appearance assumed by an electrical discharge passing through a highly rarefied gas.

4 hydrological – the existence in a lake or other body of water of two or more distinct layers differing in temperature, density or the like.

5 physical – variation in the richness of the fuel-air mixture during the period of its introduction into the cylinder of an internal combustion engine.

stratosphere
The region of the Earth's atmosphere extending above the troposphere to a height of about 50 km, in which in the lower part (up to a height of about 20 km) there is little temperature variation with height, and in the higher part the temperature increases with height. The ozone layer *(qv)* is found within the stratosphere.

strict liability (civil)
Liability which flows from the acts or omissions of the defendant even though he took all reasonable care to prevent damage being caused; sometimes called 'no-fault' liability. Of particular significance in environmental law, because of the problems that a plaintiff may face in proving negligence on the part of a defendant who has caused environmental harm, especially when that harm results from the conduct of a potentially hazardous activity. Proponents of the imposition of strict liability argue that it provides an incentive for firms to take enhanced precautions to prevent damage from occurring. Others argue that it would simply increase the cost to industry and inhibit the development of new technologies. In practice, strict liability is often mitigated by allowing the defendant to plead certain defences, such as that the damage was caused wholly by the act of some third party, or that the defendant's actions were in accord with all regulatory requirements, or that they represented the state of the art at the time they were undertaken. The European Commission's Green Paper on *Remedying Environmental Damage* (HL Paper 10; Session 1993–94) sought views about whether a strict liability regime might be introduced for environmental damage, and how it should be designed. The House of Lords Committee on the European Communities, in its observations on that Paper, accepted that strict liability might be justified in some circumstances, particularly in the case of those who knowingly engage in dangerous and potentially dangerous activities, and for past pollution by those who knew or should have known that their activities were potentially dangerous.

strict liability (criminal)

An offence for which there is no requirement that the prosecution should prove that the accused intended or could have foreseen that his acts might have resulted in the commission of an offence. As with civil liability, there is a trend towards imposing strict liability in relation to environmental offences, so as to place a heavy duty on those who have responsibility for potentially polluting activity. An example is the principal offence relating to pollution of controlled waters *(qv)* under the Water Resources Act 1991, s.85, which imposes liability on any person who causes or knowingly permits the entry of any poisonous, noxious or polluting matter to controlled waters.

The leading case is *Alphacell Ltd v Woodward* [1972] 2 All ER 475, where the House of Lords held the offence to be one of strict liability, in part because of the fear that it would not be otherwise effective to prevent river pollution. Lord Salmon (sic) observed that if the prosecution had to discharge the often impossible onus of showing that the pollution was caused intentionally or negligently, a great deal of pollution would go unpunished and undeterred, to the relief of many riparian factory owners: 'As a result, many rivers which are now filthy would become filthier still and many rivers which are now clean would lose their cleanliness'. More recent cases have continued to follow this line: see eg *Wychavon District Council v National Rivers Authority* [1993] 2 All ER 440; *National Rivers Authority v Wright Engineering Co Ltd* [1994] 4 All ER 281; *National Rivers Authority v Yorkshire Water Services Ltd* [1994] 4 All ER 274 (QBD); [1995] 1 All ER 225 (HL). See also *criminal liability.*

subsidiarity

A doctrine embedded in Article 3b of the Treaty of Rome (as amended by Maastricht) which provides:

'In areas which do not fall within its exclusive competence, the community shall take action, in accordance with the principle of subsidiarity, only if and insofar as the objectives of the proposed action can not be sufficiently achieved by the Member States and can therefore, by reason of the scale of effects of the proposed action, be better achieved by the community.'

substance

A particular kind of matter; one of a definite chemical composition. Under the Environmental Protection Act 1990, s.1(13), 'substance', as well as including natural or artificial substances, whether in solid or liquid form or in the form of a gas or vapour, is defined to include electricity or heat.

sulphur oxides

A series of oxides of the element sulphur: SO, S_2O_3, SO_2, SO_3, S_2O_7 and SO_4.

Superfund

A Federal trust fund established in the US by the Comprehensive Environmental Response, Compensation and Liability Act (CERCLA) *(qv)*, to allow the Government to act swiftly to eliminate threats to human health and minimise future risks at seriously contaminated sites. The fund is made up of contributions from Federal taxes, from a special environmental corporate tax on businesses with annual incomes over $2m, and from special taxes on petroleum and petrochemical feedstocks. However, the most controversial component of the legislation was its special liability regime under which the Government could recover its clean-up expenses from parties (known as potentially responsible parties or PRPs) who were responsible for the condition of the site. There is a wide group of potential defendants, and liability between them is both joint and several *(qv)*, and also retrospective. Defences to liability are very limited. The scheme has been widely criticised for its lack of equity and its high transaction costs and the European Commission did not propose, in its 1993 Green Paper *Remedying Environmental Damage*, adopting a similar approach to environmental liability in Europe.

Superfund Amendments and Reauthorisation Act 1986 (US) (SARA)

The 1986 reauthorisation of the original US Superfund *(qv)* legislation, which expired in September 1985. The Act represents a compromise between placing the full financial burden of clean-up of contaminated land on oil and chemical companies and a broad-based combination of business and public funding. The Superfund expired again in 1994 and the Reauthorisation Bill was not dealt with by Congress. It is likely to be reauthorised during 1995. The Act will probably include some substantive provisions including provisions dealing with lender liability *(qv)*.

surface waters

The Surface Waters (River Ecosystem) (Classification) Regulations 1994 are a component of the statutory water quality objective scheme. The Regulations prescribe a system for classifying the quality of rivers and canals, to provide the basis for setting statutory water quality objectives (SWQOs) *(qv)* under the Water Resources Act 1991 (s.83) in respect of individual stretches of water. The Regulations focus on the main aspects of water quality which are relevant to a river's ability to support fish and other forms of aquatic life. The River Ecosystem classification comprises five possible levels of river quality (RE1 being the most pure) according to concentrations of a range of specified substances.

suspended solids

Those solids which are suspended in sewage or effluent and often used as a parameter *(qv)* in water discharge consents.

sustainable development

A central concept of contemporary environmental policy. The subject of numerous definitions, but that which continues to command most widespread support was advanced by the World Commission on Environment and Development in its report *Our Common Future* (1987): 'development that meets the needs of the present without compromising the ability of future generations to meet their own needs.'

The achievement of sustainable development was a fundamental issue at the Rio Summit *(qv)* in 1992, and has been the subject of several Government reports in the United Kingdom, including *Sustaining Our Common Future* (DOE, 1989); *This Common Inheritance* Cm 1200 (1990) (and two further updating reports). Further reference: *Sustainable Development: the UK Strategy* Cm 2426 (1994); British Government Panel on Sustainable Development *(qv), 1st Report* (1995).

sustainable mobility

The key concept in the evolving transport policy of the European Commission, and the starting point for its discussion paper on *The future development of the Common Transport Policy – a global approach to the construction of a Community framework for sustainable mobility* (1011/92 (COM(92)494). The Royal Commission on Environmental Pollution, in its 18th Report, *Transport and the Environment* Cm 2674 (1994), preferred to develop its recommendations within the framework of sustainable development *(qv)*.

symbiosis

An association between organisms of different species, where both participants benefit. Examples include the pollination of flowers by insects, the relationship between leguminous plants and nitrogen-fixing bacteria, and micro-organisms in the alimentary system of ruminants and herbivores. Also known as 'mutualism'.

Tanker Owners' Voluntary Agreement concerning Liability for Oil Pollution (TOVALOP)

A voluntary industry scheme intended to be an interim solution and to remain in operation only until the compensation regime for oil pollution established by the 1967 and 1971 International Conventions had been universally adopted. Unlike the 1967 and 1971 Conventions, the voluntary schemes cover preventive measures taken if no oil is spilled and TOVALOP applies to spills from tankers in ballast.

taxonomy

The science of classification, commonly of plants and animals, including their arrangement into hierarchical groups.

technical competence

A term specifically used under the Environmental Protection Act 1990 (EPA 1990) s.74 used to describe a person who has specific and pertinent knowledge and understanding of a particular task. It does not necessarily mean the most knowledgeable person, but it is obviously one who on an assessment (by a qualified verifier or assessor) of the requirement of a particular task can be regarded to be fit and able to perform that task. The term is used in the EPA 1990, s.74(3)(b) in relation to the day-to-day control of activities authorised by a waste management licence, and in this context can be demonstrated by:

1 the possession of an appropriate certificate of technical competence (COTC) *(qv)* issued by the Waste Management Industry Training Advisory Body (WAMITAB) *(qv)* based upon the National (or Scottish) Vocational Qualification (N(S)VQ) scheme;

2 having sufficient experience to be able to be deemed technically competent under the transitional arrangements provided under the Waste Management Licensing Regulations 1994;

3 specific assessment via a structured interview by officers of the relevant waste regulation authority *(qv)*;

4 possession of other appropriate qualifications or awards.

The method chosen to demonstrate technical competence depends upon the type and size of the facility and the types of waste being handled. The management of the site must be by a technically competent person or persons.

temperature inversion

The phenomenon whereby the normal decrease of temperature with height within the troposphere is reversed, so that a warmer layer of more dense air overlies colder, less dense air. The result of the phenomenon is that rising air loses its buoyancy when meeting air at the same temperature, thereby consolidating the inversion effect. The inversion may also have the effect of trapping pollutants beneath the warm layer of air, thereby preventing their normal dispersion. This can lead to elevated levels of sulphur dioxide and particulates, and to the formation of photochemical smog, as seen in cities such as Los Angeles, where chemicals are trapped at low altitude and are exposed to sunshine. A well-documented instance of serious pollution problems related to a temperature inversion is the so-called Donora smog of October 1948. Foggy conditions over seven days in Donora, Pennsylvania, led to elevated levels of sulphur dioxide and particulates. Of the 14,000 population, 42% became ill, 10% seriously, and 18 people died.

teratogenic

Capable of causing abnormal development of the embryo and congenital malformations. Derivation *terato* = monster (Greek).

tetraethyl lead (tel)

See *lead in petrol.*

thermal efficiency

The ratio of work done by a heat engine to the mechanical equivalent of heat contained in the fuel. It is not necessarily the same as overall energy efficiency: for example, a combined heat and power *(qv)* generation scheme (CHP) may have a lower thermal efficiency than conventional power generation plant, but higher overall energy efficiency by using waste heat for local industry or housing.

The thermal efficiency of new hot water boilers using liquid and gaseous fuels is governed by EC Directive 92/42 implemented by the Boiler (Efficiency) Regulations 1993 (SI 1993 No 3083), laying down minimum efficiency requirements for new boilers.

threshold limit value (TLV)

The TLV system was developed in the US by the American Conference of Governmental and Industrial Hygienists (ACGIH). The ACGIH indices of threshold limit values and biological exposures indices for 1989–90 (Cincinnati, Ohio) define TLVs as follows:

'Threshold Limit Values refer to airborne concentrations of

substances and represent conditions under which it is believed that nearly all workers may be repeatedly exposed day after day without adverse effect.'

This does not imply that individual workers will never be affected; because of individual variation in susceptibility, some workers may suffer effects ranging from discomfort to sensitisation at exposure levels well below the TLV. The basis for setting TLVs in the US is intended to be reasonable freedom from irritation, narcosis, nuisance or impairment of health for the majority of workers. The system was adopted by the Health and Safety Executive (HSE) in the UK in its 1980 Guidance Note EH15/80, *Threshold Limit Values* (HMSO). In the mid-1980s the HSE began to publish a list of British exposure limits to chemical substances, which was subsequently updated annually: Health and Safety Executive, Guidance Note EH 40, *Occupational Exposure Limits 1984* (HMSO, 1984). This approach evolved into the statutory requirements of the Control of Substances Hazardous to Health Regulations (COSHH) 1988 (and subsequently 1994) *(qv)* as a means for determining adequate exposure by inhalation to various hazardous substances. See also *maximum exposure limit; occupational exposure standards.*

tidal power

Mechanical and electrical power produced from the rise and fall of tides. The potential for generation of tidal power has been studied for many years, particularly in relation to estuaries and tidal basins. The world's largest tidal power generation plant is situated on the estuary of the River Rance in Brittany; it was constructed in the 1960s and is capable of generating 544 kw/hours per annum. The UK Government acknowledges that tidal power offers considerable potential for the generation of renewable energy *(qv)* in the UK by means of constructing barrages with sluices and turbines across suitable estuaries. The Government's Environment White Paper (Cm 1200, September 1990) acknowledged at Annex C, para C.37, that the UK has 'unusually favourable tides' for this purpose. The Severn Estuary has the second largest tidal range in the world and has the potential to produce power to the equivalent of burning 26 million tonnes of coal per annum (*ibid*). Preliminary feasibility studies are in progress for possible tidal power schemes on the estuaries of the rivers Severn, Mersea, Loughor and Conwy. If favourable, these studies may lead to schemes being brought forward in due course.

The problems associated with tidal power schemes are firstly their very high capital cost and, secondly, the potentially serious and complex environmental impacts with their effects on sediment movement, fish migration and habitats. Under European Communities law, any scheme for a tidal power development would be likely to be categorised as an installation for hydro-electric energy falling with Annex 2 of EC Directive 85/337 on Environmental Impact Assessment *(qv)*. See also *renewable energy*.

titanium dioxide

A substance used by industry as a white pigment due to its high opacity and brilliant whiteness. It has the added advantage over earlier white pigments, such as white lead, of being potentially non-toxic and inert. It can therefore be used safely in the manufacture of products such as paint, plastic, cosmetics and food packaging. Titanium dioxide is produced from two titanium ores, limonite and rutile, both of which contain iron. The process generates acidic, iron-rich waste which has caused pollution problems. In 1972 a dispute arose between France and Italy when disposal of this waste by an Italian producer, to the Mediterranean, caused damage to the marine habitat with serious consequences for the Corsican fishing industry. It led to the development of EC Directives 78/176, 82/883, 83/29 and 89/428 to control and regulate discharges from the titanium dioxide production industry, including the three plants in the UK. 89/428 has since been replaced by EC Directive 92/112, following a successful challenge in the European Court of Justice by the European Commission as to the appropriate legal basis for the Directive: see *EC Commission v EC Council* (Case C-300/89).

The manufacture of titanium dioxide is also a prescribed process under the Environmental Protection Act 1990 (EPA 1990), and subject to integrated pollution control measures under Part 1. Further reference: National Rivers Authority, *Discharges of Waste under the EC Titanium Dioxide Directives* (Water Quality Series No 10; 1993).

toluene extractable matter (TEM)

A chemical method of analysis which is used to express the quantity of organic matter in a material. The sample is extracted with toluene and after evaporation of the toluene the residue is termed toluene extractable matter. The method is applicable to higher organic matter such as coal tar, solvents, paints and other poly-aromatic hydrocarbons. The method is commonly used for work involving contaminated land *(qv)* investigations.

tort

The breach of a civil law duty which enables a person to sue for compensation for a wrong done to his person or property. It is possible for the same wrongful act to be a crime, a breach of contract and a tort. The applicability of tort law to environmental protection stems from the remedies it affords for harm to property, principally under the heads of negligence, nuisance *(qv)* and *Rylands v Fletcher (qv)*. For an example of tort law's role in environmental protection, see *Cambridge Water Company case.*

total organic carbon (TOC)
The quantity of carbon present in the organic matter which is dissolved or suspended in water. Used as a measure of water quality.

total petroleum hydrocarbons
These are compounds of hydrogen and carbon arising from petroleum (for example from petrol and diesel), including both volatile and non-volatile compounds and particulate material. Some of these compounds are photochemically active producing ozone and other oxidants which have been associated with health effects. Others, for example methane, are relatively photochemically unreactive.

Town and Country Planning Act 1990 (TCPA 1990)
The principal legislation relating to planning control *(qv)* in England and Wales.

Town and Country Planning (Scotland) Act 1972
The principal legislation relating to planning control *(qv)* in Scotland.

Town and Country Planning (Use Classes) Order 1987
An order which specifies several classes of land use, and provides that a change of use to another use within the same class is not to be treated as development *(qv)* for the purposes of planning control *(qv)*. The relevant order for Scotland is the Town and Country Planning (Use Classes) (Scotland) Order 1989.

Toxic Release Inventory (TRI)
The US national database containing information on toxic pollution generated by manufacturing industries. Reports on the effect of the TRI collated by the US Environmental Protection Agency suggested to the US Government that annual aggregation of polluting substances from all major sources could help regulatory bodies to identify and target environmental blackspots and give the public and environmental groups information in a user-friendly form.

toxic tort
Not a technical term, but an expression in common use to refer to a

tort resulting from environmental pollution, for example damage to health or property from the discharge of waste into the environment.

trace element

Any element which occurs in a concentration of 1 mg or less in 1 kg of any substance or any element which occurs in a substance in a concentration that is less than 0·01% of the dry mass of substance. Some trace elements are essential to the body, and some are used clinically (such as lithium), whilst others can be toxic, for example mercury, lead and other heavy metals. Various analytical techniques can be used in their identification and measurement.

tradable permit

A permit under which a company which is able to abate its emissions of specified substances below the upper limit set by a regulator is entitled to sell the excess allocation to a company whose marginal abatement cost is higher than the cost to it of the excess permit allocation. Hence the scheme promotes economic efficiency by inducing abatement by the least-cost abater, rather than by imposing across-the-board restrictions. The overall bubble of allocations may be reduced progressively over time, increasing pressure on all emitters; and interest groups and others may buy emission rights so as to raise costs to industry and hence enhance the viability of abatement. An experimental scheme is presently in force in the US under the Clean Air Act Amendments 1990, in relation to sulphur dioxide emissions by electricity generating utilities.

trade effluent

Any liquid waste matter, with or without particles of matter suspended within it, which is wholly or partly produced in the course of any trade or industry carried on at trade premises, but not including domestic sewage (Water Industry Act 1991, s.141). Trade effluent may be discharged into a sewerage undertaker's public sewers only with the undertaker's consent. The sewerage undertaker is entitled to impose conditions on any consent (s.121), and to vary those conditions (though not normally within two years of granting the consent). See also *special category effluent; trade effluent agreement.*

trade effluent agreement

An agreement made under the Water Industry Act 1991, s.129, between a sewerage undertaker *(qv)* and the owner or occupier of any trade premises within their area, for the reception and disposal of trade effluents. Such an agreement may authorise discharges otherwise needing consent, but where a special category effluent *(qv)* is concerned it must be referred to the Secretary of State for his determination.

Tragedy of the Commons

The title of Hardin's classic article in 1968 which alerted public attention to the economic problems of population growth and environmental degradation. See Hardin, 'The Tragedy of the Commons' 162 *Science* 1243–1248 (1968).

transfer note

The document which accompanies the written description and identifies the waste which is to be transferred to holders in the chain of control, ie keepers, treaters, disposers, carriers, producers or brokers, under the Environmental Protection (Duty of Care) Regulations 1991 (SI 1991 No 2839). The document must state the quantity of the waste and how it is to be contained and stored, details of the transferor and transferee or name of licensing authority. It must also state the time and place of transfer of waste. The transferor and transferee must keep both the written description and the transfer note for a period of two years and supply the waste regulation authority (WRA) with a copy if required.

transfrontier shipment

The shipment of cargoes which cross national boundaries, including cargoes which contain hazardous waste which is being transported for disposal in another country. Many cargoes are moved for legal reasons, ie inadequate disposal facilities in the country of origin. However, for economic reasons some cargoes have been illegally transported to developing countries.

Under the Basle Convention *(qv)*, States may ban imports of waste, and shipments are not allowed between parties and non-parties without specific (bilateral or multilateral) agreements. Where shipments do occur, the Convention provides for a global system of environmental controls.

transpiration

The process by which water absorbed by plants, usually through the roots, is evaporated into the atmosphere from the plant surface.

transport

A major source of environmental pollution, such as a producer of noise, local and regional air pollution and greenhouse gases; and also in terms of the physical impacts of the road-building programme on habitats, homes and cities. An exhaustive study is contained in the Royal Commission on Environmental Pollution's (RCEP) 18th Report, *Transport and the Environment* (Cm 2674; 1994); see also House of Commons Transport Committee, *Transport-related Air Pollution in London* (Session 1993–94; HC 506).

transposition

The process of implementation of the requirements of European Directives, principally by incorporation into the national law of Member States.

tributyltin compounds (TBT)

Organic compounds of tin which are commonly used for preserving wood and are found in anti-fouling paint applied to the hulls of ships. Low levels of such compounds have an adverse effect on the ecology of the aquatic environment and are listed in SI 1991 No 472 (as amended) as prescribed substances *(qv)* for the purposes of integrated pollution control (IPC) *(qv)*. Any timber process, therefore, such as preserving wood, which could result in a release to water of TBT must be authorised under the Environmental Protection Act 1990.

Such is the detrimental effect of TBT on the ecology of marine waters in harbours or marinas that UK regulations prevent the use of TBT anti-fouling paints on small boats. As yet the ban does not apply to large ships.

troposphere

The lowest part of the atmosphere, which reaches from the Earth's surface up to the stratosphere, and is on average about 12 km in depth (ranging from about 11 km above polar regions to about 16 km above equatorial regions). It is the location of all weather processes. The boundary between the troposphere and the stratosphere is the tropopause.

turbidity

Opacity of liquid, due to suspended particles.

ultraviolet (UV)

Electromagnetic radiation occurring between the visible and X-ray regions of the spectrum within the wavelength range of approximately 400 nm to 10 nm. Ultraviolet radiation is normally filtered in the upper atmosphere by ozone, where it may lose energy by causing the dissociation of ozone into atomic and molecular oxygen. On the Earth's surface, exposure to certain forms of ultraviolet radiation may cause cataracts and skin cancers. Ultraviolet radiation is also involved in photochemical reactions (ie chemical reactions fuelled by light energy) and the generation of vitamin D. This type of radiation will not pass through ordinary glass windows.

underground storage tank (UST)

Typically a metallic container used to store raw materials, fuel oil, etc in bulk on site. Potentially, it is a source of contamination for groundwater and/or nearby surface waters through gradual deterioration of the metallic fabric, joints or valves. The integrity of these tanks has to be the subject of constant monitoring and indications of leakage need to be immediately investigated and rectified. See *leaking underground storage tank (LUST).*

uniform emission standard

A quantitative limit on the emission or discharge of a contaminant from a particular source, frequently an industrial operation. With a uniform emission standard the same quantitative limit is placed on all emissions of a particular contaminant. This conceptually simple system seeks to regulate an operator in relation to his operation only; it takes no account of any variation in the quality/sensitivity of environmental media into which the emissions are passed. See also *environmental quality standards.*

United Kingdom Environmental Law Association (UKELA)

A registered charity and company limited by guarantee whose main objective is to promote, for the benefit of the public generally, the enhancement and conservation of the UK environment, and to advance environmental law education. Its annual weekend University Conferences enjoy a high reputation together with the journal of the UK Environmental Law Association. Please see front pages of Lexicon for further details.

United Nations Conference on Environment and Development (UNCED)

Conference held in Rio de Janeiro in 1992 which focused on sustainable development *(qv)* and on problems in developing countries of environmental degradation, including deforestation and desertification caused by attempts to raise standards of living. UNCED succeeded in producing two important conventions, the Climate Change Convention *(qv)* and the Framework Convention on Biological Diversity, and an action plan for the next century, known as Agenda 21 *(qv)*. The plan seeks to integrate environmental concerns into a broad range of activities taking into account the different needs of developing and developed countries. See also *Rio Summit.*

United Nations Economic Commission for Europe (UNECE)

A regional committee of the Economic and Social Committee of the United Nations, and comprising representatives from 34 countries, including the Member States of the European Communities (EC),

Economic Association of Communist Countries (COMECON) and European Free Trade Area (EFTA), as well as the USA and Canada.

Its main objectives are: to encourage economic co-operation between the European Member States of the United Nations; to increase the economic strength of Europe; to encourage economic and technological co-operation, trade and environmental protection among European countries; to conduct research and studies on economic, technological and environmental problems in Europe; and to collect, analyse and distribute economic, technological and statistical data.

United Nations Environment Programme (UNEP)

A programme established in 1972 by the UN General Assembly to provide machinery for international co-operation in matters relating to the human environment. The Governing Council of UNEP, comprising 58 Member States elected by the General Assembly for four-year terms, meets every two years to review the state of the world's environment, to promote international co-operation in UNEP activities, and to provide programme policy guidance. UNEP's priorities include the protection of the atmosphere; the quality of freshwater resources; the oceans and coastal areas and resources; land resources by combating deforestation and desertification; the conservation of biological diversity; environmentally sound management of bio-technology, hazardous waste and toxic chemicals; protection of human health conditions and quality of life. A number of important programmes have been initiated by UNEP based upon these priorities, for example Global Environmental Monitoring (GEM); INFOTERRA - an international environmental information system and the International Register of Potentially Toxic Chemicals (IRPTC).

uranium

A radioactive metallic element, symbol U. Found naturally as an oxide or complex salt in various widely distributed minerals, particularly pitchblende. Used as a nuclear fuel and in nuclear weapons.

Urban Waste Water Treatment Directive (UWWTD)

EC Directive 91/271 which imposes requirements in respect of the treatment of urban waste water from urban centres of specified sizes prior to discharge to a receiving water. The UWWTD lays down standards for the discharges from sewerage works entering receiving waters (classified into sensitive waters *(qv)* and less sensitive waters *(qv)*) and an authorisation system for discharges of industrial waste water into the sewer system prior to treatment. Implementation is required by 1998, and is estimated to require capital expenditure in the United Kingdom of around £6 bn at 1993–94 prices in improvements to inland sewage treatment works and other capital investment.

Use Classes Order

See *Town and Country Planning (Use Classes) Order 1987.*

vadose zone

The zone between the top of the water table and the surface of the soil. Contamination of this zone is likely to affect adversely the growth and development of plant and bacterial life directly and animals, including man, indirectly, through food chains. Dangers to groundwater from contaminants in the vadose zone will depend upon the degree of adsorption of the contaminant onto the soil particles in the zone.

Valdez Principles

A set of 10 principles that encourage development of positive programmes designed to prevent environmental disasters and degradation. They originated as a policy, intended to reduce the environmental impact of Exxon's activities, formulated in 1989 as a corporate response to the *Exxon Valdez (qv)* Alaskan oil spill disaster. In 1990 the Coalition for Environmentally Responsible Economies (CERES) *(qv)* published its guide to the principles. CERES make the point that to halt widespread environmental damage and to create a sustainable future will require a full commitment from individuals, corporations, governments and other entities to develop positive, aggressive solutions.

Through its guide to the Valdez Principles, CERES and its signatory companies intend to help investors make informed decisions around environmental issues. As representatives of the investment and environmental communities, they ask corporations joining CERES to subscribe to these principles. The principles are: protection of the biosphere; sustainable use of natural resources; production and disposal of waste; wise use of energy; risk reduction; marketing of safe products and services; damage compensation; the use of environmental directors and managers; disclosure and assessment and annual audit.

vanadium

A metallic element, symbol V. It is used as a catalyst, and its alloys are used in applications requiring its properties of great hardness and high tensile strength. Toxic in some forms and controlled under environmental legislation, vanadium is a significant contaminant in some crude oil derived products.

vapour pressure

The pressure exerted by the vapour emitted by any substance. It is small or negligible for most solids at ordinary temperatures and pressures. Vapour pressure increases with temperature as a result of evaporation. At boiling point, the local vapour pressure equals total pressure.

vector

A carrier of disease; eg the mosquito is a vector for malaria.

vehicle emission standards

Limits on the emission or discharge of substances from car, lorry and motorcycle exhausts. Controlled by product standards regulating the emission equipment installed in cars; given effect by various Construction and Use Regulations made under the Road Traffic Act 1988.

Venturi meter

A device for measuring fluid flow by means of inserting a tapered throat into a pipe and determining the reduction in pressure at the throat.

vibration

The motion in the particles of a body creating resonance which may translate into a sound wave. Industrial operations, for example through the use of motors or machinery with moving parts, may generate vibration which could become a statutory nuisance under Part III of the Environmental Protection Act 1990 and/or a cause of action in common law nuisance.

virus

A disease-producing particle, visible to an electron microscope, which is capable of reproduction only within a living cell.

volatile organic compound (VOC)

An organic compound which evaporates easily at ambient temperatures and contributes to air pollution, mainly through the formation of secondary pollutants. VOCs include hydrocarbons (eg petrol) and other more complex organic compounds. VOCs are also implicated in stratospheric ozone depletion.

vorsorgenprinzip

A German concept deriving from the idea of forethought or anticipating the needs for tomorrow. It is a version of the precautionary principle *(qv)*. It can also mean the early detection of dangers to health and environment by comprehensive, synchronised (harmonised) research, in particular into cause and effect relationships. Its implication is that precautionary action should be taken in advance of complete understanding by science becoming available. The principle seeks to develop, in all sectors of the economy, technological processes that significantly reduce environmental burdens, especially those brought about by the introduction of harmful substances. The Royal Commission on Environmental Pollution *(qv)* 12th Report, *Best Practicable Environmental Option,* has an interesting explanation of the principle at Appendix 1 by Dr Konrad Moltke.

warranty

A term in a contract whose breach gives rise only to a claim for damages. If the term is a condition, and that condition has been breached, then that breach would give rise to the right to terminate the contract as well as a claim for damages.

waste

Discarded or unwanted materials, distinguishable from raw materials, products and useful by-products, the disposal and recovery of which are subject to a special regime of control under European Community law and under UK law (see Environmental Protection Act 1990, Part II, formerly the Control of Pollution Act 1974, Part I and, before that, the Deposit of Poisonous Waste Act 1972). Although current UK law provides no conclusive definition of the term 'waste', the Environmental Protection Act 1990, s.72(2) provides that waste includes:

1 any substance which constitutes a scrap material or an effluent or other unwanted surplus substance arising from the application of any process; and

2 any substance or article which requires to be disposed of as being broken, worn out, contaminated or otherwise spoiled.

By s.75(3) anything which is discarded or otherwise dealt with as if it were waste should be presumed to be waste unless the contrary is proved. In considering the meaning of waste under this definition (and before it under s. 30 of the Control of Pollution Act 1974, which was in almost identical terms) the courts took the view that waste was to be defined by reference to the intention of the person producing or discarding it. On that basis, the fact that material might have a use or value to the person removing or receiving it would not in itself prevent the material being waste. 'One man's waste may be another man's treasure' : *Long v Brooke* [1980] Crim. L.R. 109.

Under European Community law a definition of waste was provided by the original Waste Framework Directive *(qv)* 75/442/EEC, which defined waste as meaning: 'any substance or object which the holder disposes of or is required to dispose of pursuant to the provisions of the National Law in force.' The European Court of Justice in joined Cases C-206/88 and C-207/88 (*Vessoso & Zanetti*) held that, under this definition, substances or objects which were capable of economic reutilisation were not excluded from the definition of waste. It also held that it was not a necessary component of the definition that the holder must have intended to exclude all possibility of economic reutilisation of the material by others.

The Waste Framework Directive, as amended by EC Directive 91/156, Article 1(a) defines waste to mean 'any substance or object in the categories set out in Annex I which the holder discards or intends or is required to discard'. The 'holder' is defined to mean the producer of the waste or the person who is in possession of it; the 'producer' is defined to mean anyone whose activities produce waste and/or anyone who carries out pre-processing, mixing or other operations resulting in a change in the nature or composition of the waste. Annex I of the Directive lists some 16 categories of waste (Q1/Q16): these include production or consumption residues, off specification products, spilled, lost or contaminated materials, products for which the holder has no further use, and contaminated materials resulting from remedial action with respect to land. The final category, Q16, covers any materials, substances or products not contained in the previous 15 categories. Thus the key concept which distinguishes waste under EC law is whether the materials have been discarded by the holder, or are intended, or are required, to be discarded.

The definition of waste contained in the framework directive is incorporated into UK law by the Waste Management Licensing Regulations 1994 (SI 1994 No 1056) reg.1(3) which defines 'Directive waste' in terms of the definition in the Waste Framework Directive.

DOE Circular 11/94 provides guidance as to the meaning of waste and suggests (para 2.14) that waste appears to be perceived in the Directive as posing a threat to human health or the environment which is different from the threat posed by substances or objects which are not waste. This threat arises from the particular propensity of waste to be disposed of or recovered in ways which are potentially harmful to human health or the environment and from the fact that the producers of the substances or objects concerned will no longer have the self-interest necessary to ensure the provision of appropriate standards. This has led to the Government formulating a test for waste as being 'those substances or objects which fall out of the commercial cycle or out of the chain of utility'. The Circular discusses at length what distinguishes waste from other materials, such as by-products or fuels. Particular difficulty can be experienced in distinguishing between waste and other materials in the case of recovery operations listed under the Directive, and the Circular suggests that a helpful distinction is between the ordinary reutilisation of substances which are not waste and specialised recovery operations relating to waste, these being 'operations which are of their nature recovery operations since they wholly or partly derive their justification from the recovery of waste' (para 2.30).

The distinction between waste and non-waste materials is important, since controlled waste (*qv*) is subject to waste management licensing, to requirements for registration of carriers and to the duty of care (*qv*).

wastes, clinical
See *clinical wastes.*

waste collection authority (WCA)
The local authority *(qv)* with duties in relation to waste collection under the Environmental Protection Act 1990. It is normally a responsibility of the district council, but in Greater London it falls to the London boroughs. In Scotland, it is the islands or district council.

It is the duty of each waste collection authority to arrange for the collection of household waste in its area (with certain limited exceptions) and, if requested by the occupier of premises, to arrange for the collection of commercial waste from premises (1990 Act, s.45). It also has the power, though not the duty, to collect industrial waste from premises within its area. The authority is required to deliver for disposal all waste to such places as the waste disposal authority (*qv*) for its area may direct. Waste collection authorities also have the duty to prepare waste recycling plans under s.49 of the

Act in relation to household and commercial waste arising in their area, with a view to deciding what arrangements are appropriate for dealing with the waste by separating, baling or otherwise packaging it for recycling purposes.

waste disposal authority (WDA)

The local authority with responsibility for arranging for the disposal of controlled waste collected in its area by the waste collection authorities and providing places at which persons resident in the area may deposit their household waste (often known as civic amenity sites or recycling centres) under the Environmental Protection Act 1990 s.30(2). English county councils and Welsh district councils are the waste disposal authority. In Greater London, the WDA may be specially constituted, or in other cases will be the Common Council of the City of London, or the London Borough. In metropolitan counties in England, the metropolitan district council is the WDA, save for Greater Manchester and Merseyside, where special arrangements apply. In Scotland, the WDA is the islands or district council.

The functions of WDAs are to be fulfilled by means of arrangements made with waste disposal contractors (*qv*) but by no other means. The WDA may, however, hold land which is made available to waste disposal contractors for the purpose of enabling them to treat, keep or dispose of controlled waste; plant and equipment may similarly be held by the WDA and made available to the waste disposal contractor for the purpose of enabling him to keep such waste prior to its removal for disposal, or to treat it in connection with such keeping, or for the purpose of facilitating its transportation (for example, compacting and baling the waste). Thus in practice the WDA will discharge its statutory functions either through a local authority waste disposal company (LAWDC) (*qv*) operating at arm's length, or through contractual arrangements with a private sector waste disposal contractor.

waste disposal contractor

A person or company who, in the course of a business, collects, keeps, treats or disposes of waste, being either:

1 a company formed for all or any of these purposes by a waste disposal authority (WDA) (*qv*); or

2 a company formed for all or any of these purposes by other persons, or a partnership or an individual (Environmental Protection Act 1990, s.30 (5)).

A WDA must discharge its statutory functions of waste disposal through arrangements made with a waste disposal contractor.

Schedule 2 of the 1990 Act also makes provision relating to waste disposal contracts to be entered into between the WDA and the contractor, including the procedure for putting waste disposal contracts out to tender. In determining the terms and conditions of any contract, the WDA must avoid undue discrimination in favour of one description of waste disposal contractor as against other descriptions of waste disposal contractor (Schedule 2 para. 18). The WDA must also have regard to the desirability of including in the contract terms and conditions designed to minimise pollution of the environment or harm to human health due to the disposal of waste, and to maximise the recycling of waste (Schedule 2 para. 19(1)). These provisions were considered by the Court of Appeal in *R v Avon County Council, ex parte Terry Adams Limited* (*The Times*, January 20, 1994) where it was held that the WDA was entitled to decide which bid to prefer for the various contracts under tender, according to the honest judgment of the councillors to whom fell the task of decision, the only limitation being that the decision should be rational in the administrative law sense and should be reached in accordance with any statutory requirements.

Waste Framework Directive

EC Directive 75/442 which established a set of Community rules for waste disposal. This Directive was amended by Directive 91/156/EEC, and the two must be read together. The preamble to the amending Directive suggests that there are a number of related objectives to the Community waste policy. These include the formulation of a common terminology and definition of waste, action by Member States to ensure the responsible removal and recovery of waste, together with measures to restrict the production of waste; the encouragement of recycling of waste and reuse of waste as raw materials; the promotion of self-sufficiency at Community level and at individual Member State level for waste disposal; the formulation of waste management plans in Member States to achieve these objectives; the reduction in movements of waste; the authorisation and inspection of undertakings carrying out waste disposal and recovery; the authorisation or registration and inspection of other undertakings involved with waste, eg waste collectors, carriers and brokers, in order to monitor waste from its production to final disposal.

Article 3 of the Directive requires Member States to take appropriate measures to encourage the prevention, reduction and recovery of waste. Article 4 requires Member States to take the necessary measures to ensure that waste is recovered or disposed of without endangering human health and without using methods or processes which could harm the environment, and in particular, without risk to water, air, soil and plants and animals; without causing a nuisance through noise or odours; and without adversely affecting the countryside or places of special interest. Article 5 requires Member States to take appropriate measures, possibly in co-operation with other Member States, to establish an integrated and adequate network of disposal installations; this network should enable the Community as a whole to become self-sufficient in waste disposal and Member States can move towards that aim individually. Article 7 requires Member States, in order to attain these objectives, to draw up waste management plans, relating to types, quantity and origin of waste to be recovered or disposed of to suitable disposal sites or installations.

A permitting system for establishments or undertakings carrying out waste disposal is required by Article 9, and Article 10 requires establishments or undertakings carrying out recovery operations to obtain a permit from the competent authority. Article 12 requires establishments or undertakings which collect or transport waste on a professional basis, or which arrange for the disposal or recovery of waste on behalf of others, either to be subject to authorisation or to be registered with the competent authority. Regular inspections of such establishments or undertakings is required by Article 13.

Article 1(a) requires the Commission, acting through a committee composed of representatives of Member States, to draw up a list of wastes belonging to the categories listed in Annex I of the Directive. This 'European Waste Catalogue' *(qv)* has now been finalised.

The Directive is implemented in the UK by a variety of measures. Waste recovery and disposal operations are subject to the waste management licensing system under Part II of the Environmental Protection Act 1990 and the Waste Management Licensing Regulations 1994. Professional carriers of waste are subject to registration under the Control of Pollution (Amendment) Act 1989. Waste brokers are subject to a registration system under the Waste Management Licensing Regulations 1994 (SI 1994 No 1056). Those Regulations contain a number of exemptions from waste management licensing *(qv)*, which are intended to be consistent with the Waste Framework Directive and in particular to encourage recycling and recovery of waste; as required by the Directive, such exemptions, when carried out by establishments or undertakings, are subject to a requirement of registration with the competent authority.

Waste Management Industry Training Advisory Board (WAMITAB)

A company limited by guarantee and originally formed by the Institute of Wastes Management, the National Association of Waste Disposal Contractors and the Road Haulage Association. It now has a statutory status. WAMITAB has the responsibility of establishing the criteria and administering the system for the award of certificates of technical competence *(qv)*, which are one of the components for the test of a licence holder being a fit and proper person under s. 74 of the Environmental Protection Act 1990. By reg. 4 of the Waste Management Licensing Regulations 1994 (SI 1994 No 1056), a person is technically competent for the purposes of s. 74 of the Act in relation to a specific type of facility if, and only if, he is the holder of one of the certificates awarded by WAMITAB as being the relevant certificate of technical competence for that type of facility.

waste management licence (WML)

A licence granted by a waste regulation authority (*qv*) under the Environmental Protection Act 1990 (EPA 1990), s.35(1), authorising the treatment, keeping or disposal of any specified description of controlled waste in or on specified land, or the treatment or disposal of any specified description of controlled waste by means of specified mobile plant. A licence must be granted either to the person in occupation of the land in question or, in the case of mobile plant, to the operator of the plant. The waste management licensing system is one of the primary means through which the UK implements its obligations under the Waste Framework Directive (*qv*). Provided that planning permission or a certificate of lawful use is in place in relation to the relevant use of land, a waste regulation authority to which an application is duly made must not reject the application if satisfied that the applicant is a fit and proper person *(qv)* unless it is satisfied that rejection is necessary for the purposes of preventing:

1 pollution of the environment;

2 harm to human health; or

3 serious detriment to the amenities of the locality.

This final requirement is not applicable where planning permission for the relevant use was granted after April 30, 1994 (Waste Management Licensing Regulations 1994 (SI 1994 No 1056)).

The EPA 1990 contains provisions dealing with the variation, suspension and surrender of licences and with their transfer (ss. 37–40). Fees and charges are payable both in respect of the grant of licences and by way of an annual subsistence charge (s. 41). Further provision as to the content of licences is made by the Waste Management Licensing Regulations 1994, in particular licence conditions relating to waste oils *(qv)* and the protection of groundwater *(qv)*. Guidance as to the content and enforcement of waste management licences is contained in the Department of the Environment's series of waste management papers *(qv)*, in particular Waste Management Paper No 4 which deals with licensing in general.

waste management paper (WMP)

A series of guidance documents issued by the Department of the Environment to waste regulation authorities relating to their waste management licensing functions, and to which those authorities are legally required to 'have regard' (Environmental Protection Act 1990, s.35(8)).

Currently, WMPs are in the course of revision following the introduction of waste management licensing on May 1, 1994. Some of the papers are of a general nature, for example, dealing with the licensing of waste facilities (WMP4) and with landfill operations (WMP26A). Others deal with specialist waste streams, for example, polychlorinated biphenyl (PCB) *(qv)* wastes (WMP6), mineral oil wastes (WMP7), metal finishing wastes (WMP11) and tarry and distillation wastes (WMP13). WMP27 deals with the control and monitoring of landfill gas and WMP28 provides guidance on recycling.

waste minimisation

The minimisation of waste is an important component of EC waste policy and Article 3 of the Waste Framework Directive *(qv)* requires Member States to take appropriate measures to encourage the prevention or reduction of waste production, in particular by the development of clean technologies, by being more sparing in their use of natural resources, and by the technical development and marketing of products designed to make no contribution, or the smallest possible contribution, to increasing the amount or harmfulness of waste. The UK Government's proposed national waste strategy treats the prevention or minimisation of waste as the first stage in the preferred hierarchy of waste management. Para. 14.3 of the Government's 1990 Environment White Paper (Cm 1200) states that the Government's first priority is to reduce waste at source to a minimum, and that it intends to achieve this by imposing tougher standards on industry through integrated pollution control (IPC) *(qv)* and by promoting clean technologies. The Departments of the

Environment and of Trade and Industry run a joint environmental technology innovation scheme which is designed amongst other things to support innovatory projects in waste minimisation and new methods for treating and disposing of waste. The White Paper also states that the Government's policy is to harness market forces more effectively to encourage waste minimisation by raising standards of disposal, thereby increasing disposal costs, and, more recently, by the introduction of a proposed landfill tax. See also *landfill levy*.

waste oil

Oil arising as a waste product of the use of oils in a wide range of industrial and commercial activities, such as engineering, power generation and vehicle maintenance. During the 1970s, the improper disposal of waste oils was identified as a substantial environmental problem; a survey during this period showed that in some Member States of the European Community as much as 20%–60% of all waste oils were disposed of without any control. EC Directive 75/439 attempted to address this problem (amended by Directive 87/101). The waste oil Directives not only seek to avoid environmentally damaging means of disposal, but also to encourage the regeneration rather than the combustion of waste oils. Member States are required to give priority to regeneration 'where technical, economic and organisational constraints so allow'. The burning of waste oils which cannot be regenerated should be carried on under environmentally acceptable conditions, and waste oils which are not regenerated or burnt must be safely destroyed, or their disposal controlled. Failure to implement the waste oil Directives properly led to the issue of a Reasoned Opinion against the UK Government by the European Commission in May 1980. However, it was not until the Collection and Disposal of Waste Regulations 1988 (SI 1988 No 819) that a specific requirement for the licensing of operations processing or treating waste oil or waste solvent was introduced into UK law.

The re-refining of waste oil constitutes a recovery operation under the Waste Framework Directive *(qv)*. The use of waste oil as a fuel or otherwise as a means of generating energy likewise constitutes a recovery operation. Reg. 14 of the Waste Management Licensing Regulations 1994 (SI 1994 No 1056) requires that where a waste management licence authorises the regeneration of waste oil, it must include conditions ensuring that the base oils derived from the regeneration do not constitute a toxic or dangerous waste and do not contain polychlorinated biphenyls (PCBs) or polychlorinated terphenyls (PCTs) at all, or do not contain them in concentrations beyond a specified maximum limit which in no case is to exceed 50 parts per million. Also by reg. 14, where a waste management licence authorises the keeping of waste, it must include conditions which ensure it is not mixed with toxic and dangerous waste or with PCBs or PCTs. These provisions are necessary in order to comply with EC requirements.

The burning of waste or recovered oil in an appliance with a net rated thermal input of 3 megawatts or more constitutes a Part A process under integrated pollution control (IPC) *(qv)*: Environmental Protection (Prescribed Processes and Substances) Regulations 1991 (SI 1991 No 472), Schedule 1, Chapter 1, s. 1.3, Part A(c). The burning of waste oil or recovered oil in an appliance with a nitrated thermal input of less than 3 megawatts constitutes a Part B process under local authority air pollution control (LAAPC) *(qv)*: Environmental Protection (Prescribed Processes and Substances) Regulations 1991, Schedule 1, Chapter 1, s. 1.3, Part B(c). 'Waste oil' is defined for these purposes as any mineral-based lubricating or industrial oil which has become unfit for the use for which it was intended and, in particular, used combustion engine oil, gear box oil, mineral lubricating oil, oil for turbines and hydraulic oil. 'Recovered oil' is defined to mean waste oil which has been processed before being used.

Process Guidance Note IPR/4 (1992) deals with the combustion of waste and recovered oil in burners over 3MW (th) and contains specific initial limits relating to particulate matter, cadmium and other heavy metals, acid gases, PCBs and odour. LAAPC Process Guidance Note PG1/1 (1991) deals with waste or burners of less than 0·4 MW (th); PG1/2 (1991) deals with waste oil or recovered oil burners of less than 3 MW (th). For an interesting discussion on the different use of waste oils in other EC countries, see *House of Lords Select Committee on the EC, Disposal of Waste Oils*, 1985, HMSO. See also *waste*.

waste recycling plans

Plans which must be prepared by waste collection authorities *(qv)* for the recycling of household and commercial wastes in their area, under the Environmental Protection Act 1990, s.49. Guidance on the drafting of recycling plans, including collection of information and recycling options for local authorities, is contained in Waste Management Paper 28, *Recycling*.

waste regulation authority (WRA)

The local authority *(qv)* with the function of waste management licensing under the Environmental Protection Act 1990. For non-metropolitan counties in England it is the county council. For Greater London, Greater Manchester and Merseyside, specifically constituted authorities are the Waste Regulation Authority, ie the London Waste Regulation Authority, the Greater Manchester Waste Disposal Authority and the Merseyside Waste Disposal Authority. In other metropolitan counties, the metropolitan district council is the WRA. In Wales, it is the district councils and in Scotland the islands or district council. The Secretary of State has power to establish joint regional WRAs if it appears this would be advantageous. WRAs have the function of granting waste management licences *(qv)*, and are also under a duty to prepare waste disposal plans of the arrangements made or proposed to be made for the treatment or disposal of waste within their area (EPA 1990, s. 50). WRAs also have the function of registering waste brokers, and carriers of waste, and supervising the requirements of the Special Waste Regulations 1980 as to consignment notes *(qv)* for special waste.

Waste Strategy for England and Wales

The Government issued in January 1995 a draft Waste Strategy, which attempts to apply the principles of sustainable development. It explains how the public sector as well as individual householders and consumers can contribute to managing the waste of England and Wales in a more sustainable fashion. Priority in the Strategy is given to waste reduction, reuse, recycling, composting and land-spreading, energy from waste and then disposal to landfill, incineration, other disposal methods and permanent storage:

Some of the principal targets and action points are to:

- stabilise the production of household wastes at its 1995 level
- reduce the proportion of controlled waste going to landfill by 10% over the next 10 years with a similar reduction in the next 10 years (2015 AD)
- recycle 25% of household waste by 2000
- increase use of recycled waste materials as aggregates in England from 30 mtpa to 55 mtpa by 2006
- encourage 75% of companies of more than 200 employees to publish waste environmental policies by the end of 1999

water balance

The process of calculation and assessment of leachate *(qv)* production in a landfill *(qv)*. Its importance lies in the control of leachate production in a landfill by ensuring liquid inputs do not exceed available absorptive capacity of the waste or the capacity of control measures required to deal with that leachate produced. As such it is fundamental to the protection of water quality. Factors contributing to the water balance of a landfill include water input (ie rainfall, surface and groundwater *(qv)* infiltration and liquid waste deposit); exposed surface area, nature of wastes (including their absorptive capacity); geology of the site and available surface liquid storage capacity; and any leakage between the site and surface water or groundwater. Seasonal adjustment to water balance equations is recommended in temperate climates due to the wide seasonal variations in precipitation that are commonly experienced.

For the normal purpose of estimating generation of leachate at an operational landfill the following water balance equation may be used:

$$Q = I - E - aW$$

where:

Q = Free leachate generated at the site (m^3/annum)
I = Total liquid input (m^3/annum)
E = Evapotransparative (m^3/annum)
A = Absorptive capacity of waste (m^3 tonne waste as received)
W = Weight of waste deposited (tonnes/annum)

Leachate control systems in containment sites have to be particularly well designed to account for leachate produced as it is likely they will be the only route for removal of leachate, leachate retention will be high in such cases and thus water balance calculations will be vital for control system design.

Water Industry Act 1991 (WIA 1991)
One of the four main Acts resulting from the 1991 consolidation of water legislation, and concerned with the provision of water supplies and sewage treatment including the powers of the water and sewerage undertakers and the Office of Water Services (OFWAT) *(qv)*.

water protection zone
A zone designated by the Secretary of State under the Water Resources Act 1991, s.93, with a view to preventing or controlling pollution there, or to prohibit or restrict potentially polluting activities in that area (although it does not include nitrate from agricultural sources: see *nitrate sensitive areas*).

water quality objective (WQO)
Target for water quality in a defined stretch of water, and normally

expressed in qualitative terms, such as capacity to sustain fisheries or for abstraction of drinking water. WQOs have for many years been set and applied on a non-statutory basis, largely based on a classification system developed by the former National Water Council with a five-fold river quality classification. These are now being converted into statutory WQOs under the Water Resources Act 1991, ss.82–84. The first step in a broader approach is the classification of waters for which quality objectives are to be prescribed, for which the Secretary of State has made the Surface Waters (River Ecosystem) (Classification) Regulations 1994 (SI 1994 No 1057) specifying the classification scheme to be used when setting WQOs for the River Ecosystem Use in rivers and watercourses in England and Wales. They require the National Rivers Authority (NRA) *(qv)* to assess compliance with the WQOs in accordance with prescribed procedures and principles, including sampling methods. The Regulations specify the general purposes for which the waters in each class are to be suitable (WQOs) and the specific requirement as to concentrations of substances allowed or required (water quality standards *(qv)*). Once set, WQOs may be reviewed, normally only after five years; but it is the duty of both the Secretary of State and the NRA to exercise their water pollution powers to ensure so far as they can so do that the WQOs specified for any waters are achieved at all times (s.84). It is expected that discharge consents will be the major mechanism for attainment of statutory quality objectives, although these do not apply to non-point source pollution. Further reference: NRA Water Quality Series No 17, *Discharge Consents and Compliance* (1994).

water quality standard

An environmental quality standard *(qv)* in relation to water, normally a quantitative standard which translates the targets of a water quality objective (WQO) *(qv)* into measurable parameters in classified waters, such as a definition of the maximum concentration of contaminating substances.

Water Resources Act 1991 (WRA 1991)

One of the four main Acts resulting from the 1991 consolidation of water legislation, and concerned with the control of water resources (including licensing of abstraction and discharge), the setting of water quality objectives (WQOs) *(qv)* with flood defences and other miscellaneous functions.

Water Services Association

An association of the water companies in England and Wales (previously the Water Authorities Association). The Water Services Association acts as an intelligence-gathering and opinion-forming body and also as a press and public relations organisation which lobbies on behalf of the water services companies.

watershed

The elevated boundary line separating the head streams which are tributaries to different river systems or catchment basins.

water table

The surface of the groundwater which is the upper surface of a zone of saturation, save where that surface is formed by an impermeable layer or strata. The surface is uneven and variable and tends to rise during wet weather. The lowest level to which a water table falls in any locality is known as the permanent water table.

water undertaker

A company holding appointment under the Water Industry Act 1991 to provide a public water supply in a specified area. Some of the appointed companies also hold appointment under the same legislation as sewerage undertakers *(qv)*; others are statutory water companies with no sewerage functions. Further reference: Department of the Environment (DOE) and Welsh Office (WO), *Instrument of Appointment of the Water Undertakers* (HMSO, 1989).

well

A shaft sunk in the ground for the abstraction of water, oil or other subterranean resource. Usually of large diameter and often lined with brick or concrete, and normally of shallow depth. Different from a borehole *(qv)*, which is usually of narrow diameter, of greater depth and mechanically drilled.

wetland

An area of 'marsh, fen, peatland or water, whether natural or artificial, permanent or temporary, with water that is static or flowing, fresh, brackish or salt, including areas of marine water, the depth of which does not exceed 6 metres' (the Ramsar Convention *(qv)* on Wetlands of International Importance).

Wheal Jane

A disused Cornish tin mine which was closed in 1991, whereupon internal pumping operations, which had previously depressed the water table by 400 metres, ceased. As the water level in the mine rose, over 10 million gallons of heavily contaminated water burst from the mine in January 1992, causing serious pollution of the Carnon River from cadmium, zinc, arsenic and iron. A temporary treatment operation was arranged by the National Rivers Authority, but because of the special exemption from liability for pollution from abandoned mines (see also *mines, abandoned*), proceedings were not instituted against the owners.

wholesomeness

A requirement of any water supplied by a water undertaker *(qv)* to any premises for domestic or food production purposes, under the Water Industry Act 1991, s.68. Wholesomeness is to be construed in accordance with Regulations made by the Secretary of State (s.93(1)), and the current Regulations, the Water Supply (Water Quality) Regulations 1989 (SI 1989 No 1147), transpose the requirements of EC Directive 80/778 on drinking water. Water supplied for domestic purposes is regarded as wholesome if:

1 it meets the standards prescribed in the Regulations for particular properties, elements, organisms or substances;

2 the hardness or alkalinity of water which has been softened or desalinated is not below the prescribed standards; and

3 it does not contain any element, organism or substance, whether alone or in combination, at a concentration value which would be detrimental to public health.

Wild Birds Directive

See *Birds Directive*.

wind energy

A form of renewable energy generated by harnessing wind power, using turbines which may either generate mechanical power for pumping water or generate electricity. The UK Government believes that electricity generation from wind turbines is on the verge of widespread exploitation, and that the UK's windy climate makes it well placed to succeed in this venture. Detailed advice on wind energy and its implications for land-use planning is contained in the special annex to PPG22 *Renewable Energy* (1993).

World Conservation Monitoring Centre

A non-profit organisation funded by the International Union for the Conservation of Nature (IUCN) *(qv)*, the World Wide Fund for Nature (WWF) *(qv)* and the United Nations Environment Programme (UNEP *(qv)*. The Centre maintains an authoritative bank of data on global conservation issues, and aims to support conservation and sustainable development by providing comprehensive and up-to-date information and technical services based on research and analysis. [*Address: 219 Huntingdon Road, Cambridge, CB3 0DL; T: 01223 277314; F: 01223 277136.*]

World Health Organisation (WHO)

A United Nations specialised agency greatly involved in setting environmental standards, eg toxicological standards for pesticides and chemicals in water. It is based in Geneva and Manila and has worldwide influence.

World Heritage Site

A site identified under the 1972 UNESCO World Heritage Convention, which provides for the identification, protection, conservation and presentation of cultural and natural sites of outstanding universal value. There are presently 10 designated sites in England, and their protection is principally through planning control.

World Wide Fund for Nature (WWF)

Previously called the World Wildlife Fund, this is a non-governmental organisation aiming to achieve conservation of natural resources globally.

X-ray

Penetrating electromagnetic radiation of short wavelength, of a higher frequency than visible light but lower than gamma radiation. X-rays are usually generated by the acceleration of electrons to a high velocity using an electrostatic field in a vacuum tube and stopping them suddenly by collision with a solid body. X-rays can also be generated by inner shell transitions of atoms with an atomic number greater than 10.

yeasts

The collective term for a number of unicellular fungi from several classes, whose common characteristics are related to their structure and mode of reproduction. The order Endomycetales includes the family Saccharomycetaceae (the yeast) containing commercially important species such as *Saccharomyces cerevisiae* (brewers yeast) and *S.ellipsoideus* used for wine making). Imperfect yeasts are effectively fungi of a range of species which can attain a yeastlike state rather than their normal filamentatious form, such as the parasitic genus *Candida*, which causes thrush in man.

zinc

A bluish-white lustrous metal, ductile when pure, of group IIb of the periodic table. It has an atomic number of 30, an atomic weight of 65·37, specific gravity of 1·133 (25° C), melting point of 420° C, boiling point of 907° C, is explosive as a powder, insoluble in water, soluble in acids and alkalis and is strongly electropositive. Zinc has extensive uses as a component of many alloys and within the electrical, paint and plating industries. It is also considered an essential element for growth in animals and humans but can prove toxic to many plants when present in the soil in soluble form. It is normally found and mined in the form of sulphide, carbonate silicate or in a combined ore with manganese and iron oxide. In its natural state it contains five stable isotopes but 10 other unstable isomers and nuclides are recognised. See also *phytotoxic*.

UKELA 1995

SCHEDULE OF SELECTED EC DIRECTIVES AND REGULATIONS AFFECTING THE ENVIRONMENT

75/439	on disposal of waste oils
75/440	concerning quality required of surface water intended for abstraction of drinking water in Member States
75/442	on the disposal of waste (see 91/156 below)
76/403	on polychlorinated biphenyls and polychlorinated terphenyls
76/464	on pollution caused by certain dangerous substances discharged into the aquatic environment of the community
78/631	on approximation of laws of Member States relating to classification, packaging and labelling of dangerous preparations (pesticides)
78/659	on quality of freshwaters needing protection or improvement in order to support fish life
79/409	on the conservation of wild birds
80/68	on protection of groundwater against pollution caused by certain dangerous substances
80/778	relating to quality of water intended for human consumption
80/779	on air quality limit values and guide values for sulphur dioxide and suspended particulates
82/501	on major accident hazards of certain industrial activities
82/883	on procedures for surveillance and monitoring of environments concerned by waste from titanium dioxide industry
82/884	on limit values for lead in air
83/513	on limit values and quality objectives for cadmium discharges
84/156	on limit values and quality objectives for mercury discharges by sectors other than chlor-alkali industry
84/360	on combating of air pollution from industrial plants
84/491	on limit values and quality objectives for discharges of hexachloro-cyclohexane
84/532	on approximation of laws of Member States relating to common provisions for construction plant equipment
84/631	on supervision and control within the EC of transfrontier shipment of hazardous waste (superseded by 259/93 see below)
85/337	on assessment of effects of certain public and private projects on the environment

85/339	*on containers of liquid for human consumption* repealed by 94/62
86/278	on protection of the environment and in particular of soil, when sewage sludge is used in agriculture
86/280	on limit values and quality objectives for discharges of certain dangerous substances included in list I of annex to 76/464.
87/217	on prevention and reduction of environmental pollution by asbestos
88/379	approximation of laws, regulations and administrative provisions of Member States relating to classification, packaging and labelling of dangerous preparations
89/369	on prevention of air pollution from new municipal waste incineration plants
89/429	on reduction of air pollution from existing municipal waste incineration plants
90/313	on freedom of access to information on the environment
91/156	amending directive 75/442 on waste
91/157	on batteries and accumulators containing certain dangerous substances
91/271	concerning urban waste water treatment
91/594	regulation on substances that deplete the ozone layer
91/676	concerning protection of waters against pollution caused by nitrates from agricultural sources
91/689	on hazardous waste
91/692	standardising and rationalising reports on implementation of certain directives relating to the environment
92/32	amending for the 7th time directive 67/548 on the approximation of laws, regulations and administrative provisions relating to classification, packaging and labelling of dangerous substances
92/43	on the conservation of natural habitats and of wild fauna and flora
92/72	on air pollution by ozone
92/112	on procedures for harmonising programmes for reduction and eventual elimination of pollution caused by waste from the titanium dioxide industry
92/2455	regulation concerning export and import of certain dangerous chemicals
93/67	laying down the principles for assessment of risks to man and the environment of substances notified in accordance with 67/548
93/259	regulation on supervision and control of shipments of wastes within, into and out of the EC
93/793	regulation on the evaluation and control of the risk of existing substances
94/3	decision establishing a list of wastes pursuant to article 1(a) of 75/442 on waste
94/62	on packaging and packaging waste
94/1488	regulation laying down the principles for the assessment of risks to man and the environment of existing substances in accordance with council regulation 793/93